Lecture Notes in Control and Information Sciences

Edited by A. V. Balakrishnan and M. Thoma

Vol. 1: Distributed Parameter Systems: Modelling and Identification
Proceedings of the IFIP Working Conference, Rome, Italy, June 21–26, 1976
Edited by A. Ruberti
V, 458 pages. 1978

Vol. 2: New Trends in Systems Analysis
International Symposium, Versailles, December 13–17, 1976
Edited by A. Bensoussan and J. L. Lions
VII, 759 pages. 1977

Vol. 3: Differential Games and Applications
Proceedings of a Workshop, Enschede, Netherlands, March 16–25, 1977
Edited by P. Hagedorn, H. W. Knobloch, and G. J. Olsder
XII, 236 pages. 1977

Vol. 4: M. A. Crane, A. J. Lemoine
An Introduction to the Regenerative Method for Simulation Analysis
VII, 111 pages. 1977

Vol. 5: David J. Clements, Brian D. O. Anderson
Singular Optimal Control: The Linear Quadratic Problem
V, 93 pages. 1978

Vol. 6: Optimization Techniques
Proceedings of the 8th IFIP Conference on Optimization Techniques, Würzburg, September 5–9, 1977
Part 1
Edited by J. Stoer
XIII, 528 pages. 1978

Vol. 7: Optimization Techniques
Proceedings of the 8th IFIP Conference on Optimization Techniques, Würzburg, September 5–9, 1977
Part 2
Edited by J. Stoer
XIII, 512 pages. 1978

Vol. 8: R. F. Curtain, A. J. Pritchard
Infinite Dimensional Linear Systems Theory
VII, 298 pages. 1978

Vol. 9: Y. M. El-Fattah, C. Foulard
Learning Systems:
Decision, Simulation, and Control
VII, 119 pages. 1978

Vol. 10: J. M. Maciejowski
The Modelling of Systems with Small Observation Sets
VII, 241 pages. 1978

Vol. 11: Y. Sewaragi, T. Soeda, S. Omatu
Modelling, Estimation, and Their Applications for Distributed Parameter Systems
VI, 269 pages. 1978

Vol. 12: I. Postlethwaite, A. G. J. McFarlane
A Complex Variable Approach to the Analysis of Linear Multivariable Feedback Systems
IV, 177 pages. 1979

Vol. 13: E. D. Sontag
Polynomial Response Maps
VIII, 168 pages. 1979

Vol. 14: International Symposium on Systems Optimization and Analysis
Rocquentcourt, December 11–13, 1978;
IRIA LABORIA
Edited by A. Bensoussan and J. Lions
VIII, 332 pages. 1979

Vol. 15: Semi-Infinite Programming
Proceedings of a Workshop, Bad Honnef, August 30 – September 1, 1978
V, 180 pages. 1979

Vol. 16: Stochastic Control Theory and Stochastic Differential Systems
Proceedings of a Workshop of the „Sonderforschungsbereich 72 der Deutschen Forschungs gemeinschaft an der Universität Bonn"
which took place in January 1979 at Bad Honnef
VIII, 615 pages. 1979

Vol. 17: O. I. Franksen, P. Falster, F. J. Evans
Qualitative Aspects of Large Scale Systems
Developing Design Rules Using APL
XII, 119 pages. 1979

Vol. 18: Modelling and Optimization of Complex Systems
Proceedings of the IFIP-TC 7 Working Conferenc Novosibirsk, USSR, 3–9 July, 1978
Edited by G. I. Marchuk
VI, 293 pages. 1979

Vol. 19: Global and Large Scale System Models
Proceedings of the Center for Advanced Studies (C
International Summer Seminar
Dubrovnik, Yugoslavia, August 21–26, 1978
Edited by B. Lazarević
VIII, 232 pages. 1979

Vol. 20: B. Egardt
Stability of Adaptive Controllers
V, 158 pages. 1979

Vol. 21: Martin B. Zarrop
Optimal Experiment Design for Dynamic System Identification
X, 197 pages. 1979

For further listing of published volumes please turn over to inside of back cover.

Lecture Notes in Control and Information Sciences

Edited by A.V. Balakrishnan and M. Thoma

41

M. Gössel

Nonlinear Time-discrete Systems

A General Approach by Nonlinear Superposition

Springer-Verlag Berlin Heidelberg GmbH 1982

Author
Dr. sc. nat. M. Gössel
Akademie der Wissenschaften der DDR
Zentralinstitut für Kybernetik und Informationsprozesse
DDR-1080 Berlin
Kurstraße 33

With 36 Figures

Licensed edition for
Springer-Verlag Berlin Heidelberg New York

With exclusive distribution rights for all nonsocialistic countries:
Springer-Verlag Berlin Heidelberg New York

ISBN 978-3-540-11914-2 ISBN 978-3-540-39510-2 (eBook)
DOI 10.1007/978-3-540-39510-2

Since bulky books are not a pleasure

Francois Villon

Preface

Linear systems theory is widely applied in different areas of natural science and engineering, although many of the systems under consideration are in reality nonlinear ones.

The reason for this fact is that linear systems are relatively easy to understand and to handle.

Thus it is well-known that the input-output behaviour of linear systems can be described by means of impulse responses. Therefore linear systems can be identified by identifying only their impulse responses, and the explicit state-dependent modelling of linear systems is relatively simple if the impulse responses are known. Furthermore an appropriate linear combination of two solutions of a linear ordinary or partial-differential equation is again a solution of the differential equation under consideration, i. e. new solutions can be obtained from known ones by a simple linear combination. Thus the question arises how to extend the theory of linear systems to nonlinear systems in such a way that the advantages of linear systems theory will be conserved (in a modified form) for nonlinear systems, too.

The crucial point is that linear systems satisfy a superposition principle with respect to the operation addition (or some linear functions). Moreover, linear systems can be characterized as linear superponable systems (s. e. g. /46/).

Thus it seems to be natural to generalize the theory of linear systems by generalizing the superposition principle as it will be done in this book.

Linear systems are usually described by the fact that their systems equations are linear. On the other hand nonlinear systems have been described till now by the lack of such a definite property only.

The determination of appropriate subclasses of nonlinear systems has to be considered as an essential first step in the elaboration of a theory of nonlinear systems.

In this book the different subclasses of nonlinear systems are defined by their different superposition properties and **the uniform theory of nonlinear systems is proposed as a theory of nonlinear superposition.**

Mainly we are interested in time-discrete systems, although time-continous systems are also considered.

As a first step of the generalization of the linear superposition principle we investigate the superposition of an automaton $\mathfrak{a}$ with respect to a pair $(\circ, \nabla)$ of 2-ary operations defined on the input- and output set of $\mathcal{A}$, respectively.

In a second step of generalization the 2-ary operations o and $\triangledown$ (without memory) are replaced by automata $\mathcal{C}$ and $\mathcal{D}$ with two input- and one output terminals each. It is shown that for every automaton $\mathcal{A}$ there exists a pair of automata ($\mathcal{C}$, $\mathcal{D}$) such that $\mathcal{A}$ is superponable with respect to ($\mathcal{C}$, $\mathcal{D}$). This fact allows us to classify all (nonlinear!) automata by means of their superposition properties.

Furthermore the concept of invariant relations for automata is presented. As a natural generalization of the superposition with respect to operations this concept enables us to extend the description of the input-output-behaviour by means of generalized impulse responses to automata preserving a pàir of relations.

For the case of time continous systems we describe and discuss different conditions under which the nonlinear superpositions of solutions of ordinary and partial differential equations are again a solution of the considered differential equations.

Of course the theory of nonlinear systems as a theory of nonlinear superponable systems is at the very beginning. Only some first results are presented here.

This book consists of 5 chapters. Chapter 1 is a brief review of the basic notions and notations of automata theory.

Chapter 2 deals with linear automata. The close connection between linearity and linear superponability is explained and the description of the input-output-behaviour by means of impulse responses is discussed in detail.

In chapter 3 the generalized superponability with respect to a pair of operations (o,$\triangledown$) is investigated. It is shown that (o,$\triangledown$)-superponable automata can be characterized by two operations of generalized bisymmetry. For different assumptions concerning the operations o and $\triangledown$ different classes of superponable automata are characterized and investigated. Thus binary superponable automata, (o,$\triangledown$)-linear automata, automata superponable with respect to rational operations, quasiadditions and others are considered in detail.

Chapter 4 is dedicated to the superponability of an automaton $\mathcal{A}$ with respect to a pair of automata ($\mathcal{C}$, $\mathcal{D}$). It is shown that every automaton $\mathcal{A}$ is ($\mathcal{C}$, $\mathcal{D}$)-superponable for an appropriate choice of $\mathcal{C}$ and $\mathcal{D}$. For the special case that $\mathcal{A}$ is invertible, the set of all pairs of automata for which $\mathcal{A}$ is ($\mathcal{C}$, $\mathcal{D}$)-superponable is characterized.

Chapter 5 deals with the concept of invariant relations for automata. It is shown how linear automata, superponable automata, homomorphisms and tolerance automata can be considered from the point of view of invariant pairs of relations. As we have already mentioned, this concept allows us to extend the possibility to describe the input-output-behaviour of a large class of nonlinear automata preserving a pair of relations by means of generalized impulse responses.

Appendix I is a brief report on so-called homomorphic systems. Homomorphic systems are such (nonlinear) systems that can be interpreted as algebraically linear transformations of vector spaces under an appropriate definition of vector addition and scalar multiplication. This appendix is based on /31/, which is the first paper on nonlinear superposition.

In appendix II known results concerning the superposition of solutions of ordinary and some partial differential equations are presented. The notion (nonlinear) "connecting function of a differential equation" is explained. Connecting functions of some classes of differential equations are discussed.

Acknowledgements. A lot of the ideas presented here arose in a long cooperation with Dr. sc. R. Pöschel from the Institute of Mathematics of the Acad. of Sciences of the GDR, Berlin.

I am very grateful to Prof. Dr. sc. V. Kempe, Director of the Institute of Cybernetics and Information Processing and Prof. Dr. M. Peschel, Head of the Scientific Department Mathematics/Cybernetics for their interest, for stimulating discussions and especially for the possibility to work in this field for a long time.

I thank Dr. R. Höppner from the Akademie-Verlag Berlin for his friendly and effective cooperation.

I also thank Dozent Dr. W. Wechler, TU Dresden, for critical remarks.

Contents

1. Automata - definitions and notations

In this chapter we state some standard concepts of automata theory that are used in this book. This is intended as a short review and not as a true introduction.

Like other theories motivated by needs of natural science and engineering, automata theory is concerned with mathematical models which serve as approximations to physical, technical, biological, or abstract phenomena. The model abstract automaton is the prototyp of a time discrete system. Within the concept of automata theory a system is described by mathematical laws which determine its observable behaviour and not by its special internal physical structure. This is the reason why automata theory is useful in a wide variety of fields.

In this book we consider completely defined initial deterministic automata $\mathfrak{a} = (X,Y,Z,z_o,\delta,\lambda)$ where the input-, output- and state-sets are X, Y and Z, respectively. The initial state is z_o, $z_o \in Z$, the next-state function is $\delta : Z \times X \rightarrow Z$, and the output-function is $\lambda : Z \times X \rightarrow Y$. We suppose $\mathfrak{a}$ to be synchronous and time-discrete. Let t, t = 0,1,2,... denote the discrete time parameter. Then we put

$$z_{t+1} = \delta(z_t,x_t) \tag{1}$$

$$y_t = \lambda(z_t,x_t). \tag{2}$$

The next state z_{t+1} at time t+1 is uniquely determined by the present state z_t and the present input x_t at time t. The output y_t is a function of the present state z_t and the present input x_t. As usual we extend the next state and the output functions by the formulae

$$\begin{aligned} \delta(z,\Lambda) &= z \\ \delta(z,px) &= \delta(\delta(z,p),x) \end{aligned} \tag{3}$$

and

$$\begin{aligned} \lambda(z,\Lambda) &= \Lambda \\ \lambda(z,px) &= \lambda(z,p)\ \lambda(\delta(z,p),x) \end{aligned} \tag{4}$$

for each $x \in X$, $z \in Z$ and $p \in X^*$, where X^* is the set of all finite words of elements of X, and Λ is the empty word.

The extended next-state function $\delta : Z \times X^* \rightarrow Z$ and the extended output function $\lambda : Z \times X^* \rightarrow Y$ are denoted by δ and λ, respectively, too. We put $\lambda(z_o,p) = \lambda(p)$ for $p \in X^*$ if z_o is the initial state of the automaton under consideration.

The automaton $\mathfrak{a}$ is called a Mealy-type automaton. If we have, instead of Eq. (2),

$$y_t = \mu(z_t), \tag{5}$$

then $\mathfrak{a}$ is called a Moore-type automaton. For a Moore-automaton the output y_t at time t depends on the state z_t at time t only. A state automaton (Medwedjew-automaton) $\mathfrak{a} = (X,Z,z_o,\delta)$ can be interpreted as a Moore-automaton with $\mu(z) = z$ for $z \in Z$. (X is the input set, Z is the state set, z_o is the initial state and $\delta : Z \times X \to Z$ is the state transition function.)
For a given automaton $\mathfrak{a}$ we define

$$g_t^{\mathfrak{a}} (x_1,\ldots,x_t) = \lambda(\delta(z_o,x_1,\ldots,x_{t-1}),x_t), \tag{6}$$

$y_t = g_t^{\mathfrak{a}} (x_1,\ldots,x_t)$ is the output of $\mathfrak{a}$ at time t in response to the input word $x_1 \ldots x_t$.
For a given automaton $\mathfrak{a}$ the sequence of functions $g_t^{\mathfrak{a}} : X^t \to Y$, $t = 1,2,3,\ldots$ is uniquely determined and each sequence $g_t : X^t \to Y$, $t = 1,2,3,\ldots$ determines (up to equivalence) an automaton $\mathfrak{a}$ with $g_t^{\mathfrak{a}} = g_t$.
For $p = x_1 \ldots x_t$ we sometimes write $g_t^{\mathfrak{a}} (x_1,\ldots,x_t) = g_t^{\mathfrak{a}} (p)$.
Y_z denotes the set of all possible outputs of $\mathfrak{a}$ in state z.

$$Y_z = \{y \,/\, \exists\, x \in X,\ \lambda(z,x) = y\}. \tag{7}$$

A function $g_t : X^t \to Y$ shall be called essential if it actually depends on at least two variables (i. e. at least two variables are not fictitious).

An automaton $\mathfrak{a}$ is called essential if the function $g_t^{\mathfrak{a}}$ is essential for at least one t, $t > 0$.

In the following we are mainly interested in essential functions and automata.
For $k > 0$ Y_k denotes the set of possible outputs of $\mathfrak{a}$ at time $t = k$.

$$Y_k = \{y \,/\, \exists\, p \in X^k,\ y = g_k^{\mathfrak{a}}(p)\}. \tag{9}$$

An automaton $\mathfrak{a} = (X,Y,Z,z_o,\delta,\lambda)$ is called finite if X, Y and Z are finite.

There are different possibilities of representation of automata.

1. The functions $\delta : Z \times X \to Z$ and $\lambda : Z \times X \to Y$ may be represented in a tabular form referred to as the transition table.

For all possible ordered pairs (z,x), $z \in Z$, $x \in X$, the values of the two functions δ and λ are listed in these tables.

2. An automaton may be skeletonized as a directed graph known as the state diagram (or state graph). The vertices and the directed arcs of the graph correspond to the states and the state transitions, respectively. The directed arcs are drawn and labelled with input and output symbols according to the following rule: A directed arc is drawn from the vertex labelled z to the vertex labelled z' with the arc label x,y if $\delta(z,x) = z'$ and $\lambda(z,x) = y$.
Other methods of representation of automata are not needed within this book.
As a simple example let us consider the automaton $\mathfrak{A} = (X,Y,Z,z_o,\delta,\lambda)$ with $X = \{1,2,3\}$, $Y = \{a,b\}$, $Z = \{A,B\}$ with δ and λ in tabular form:

x \ z	A	B
1	B	A
2	A	A
3	B	B

Table of the next state function

x \ z	A	B
1	a	b
2	a	b
3	b	a

Table of the output function

The corresponding state diagram of $\mathfrak{A}$ is given in Fig. 1.

Fig.1

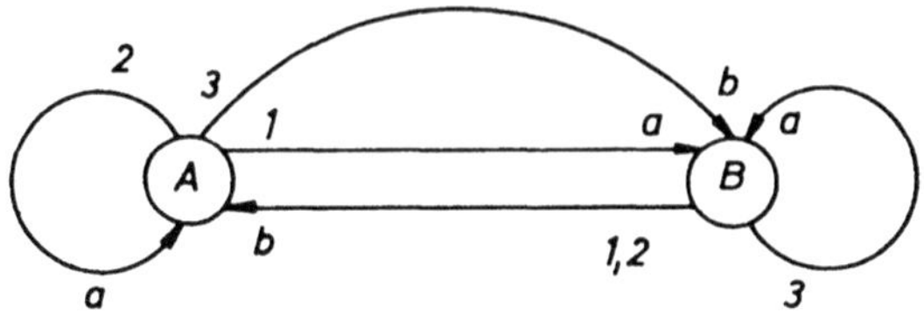

Of course, both the state diagram and the transition table contain the same information about $\mathfrak{A}$, and the one can be obtained directly from the other.

A state z' is called reachable from z if there exists a word $p \in X^*$ such that $z' = \delta(z,p)$. The set of states reachable from the initial state z_o will be denoted by Z_r,

$$Z_r = \{z \,/\, z \in Z,\ \exists\, p \in X^* \wedge z = \delta(z_o,p)\}. \tag{9}$$

The concept of equivalence plays an important role in automata theory.

Equivalent states are states with the same input-output-behaviour.

Let $\mathfrak{A} = (X,Y,Z,z_0,\delta,\lambda)$ and $\mathfrak{A}' = (X,Y,Z',z_0',\delta',\lambda')$ be automata. The state $z \in Z$ (of the automaton $\mathfrak{A}$) and the state $z' \in Z'$ (of the automaton $\mathfrak{A}'$) are said to be equivalent, $z \sim z'$, if and only if for all $p \in X^*$

$$\lambda(z,p) = \lambda'(z',p). \tag{10}$$

Thus z and z' are equivalent if and only if there is no way of distinguishing by observing the input-output-behaviour between the automaton $\mathfrak{A}$ in initial state z and the automaton $\mathfrak{A}'$ in initial state z'.

This definition of the equivalence of states also applies to the special case where $\mathfrak{A} = \mathfrak{A}'$.

It is easy to verify that state equivalence is an ordinary equivalence relation.

Furthermore the equivalence of automata can be defined.

Let $\mathfrak{A} = (X,Y,Z,\delta,\lambda)$ and $\mathfrak{A}' = (X,Y,Z',\delta',\lambda')$ be automata (without specified initial states). Then $\mathfrak{A}$ and $\mathfrak{A}'$ are said to be equivalent if for every state $z_1 \in Z$ there exists a state $z_1' \in Z'$ with $z_1 \sim z_1'$ and if for every state $z_2' \in Z'$ there exists a state $z_2 \in Z$ with $z_2 \sim Z_2'$.

If $\mathfrak{A}$ and $\mathfrak{A}'$ are initial automata with the initial states z_0 and z_0', respectively, and if $z_0 \sim z_0'$, then $\mathfrak{A}$ and $\mathfrak{A}'$ are called initially equivalent.

An automaton which does not possess mutually equivalent states is called reduced or minimal. More precisely:

The automaton $\mathfrak{A}$ is called reduced or minimal if $z_1 \sim z_2$, $z_1, z_2 \in Z$ implies $z_1 = z_2$.

Sometimes we need the equivalence of states with respect to a subset $\tilde{X}$ of X.

Let $\tilde{X} \subseteq X$. Then $z \in Z$, $z' \in Z'$ are called $\tilde{X}$-equivalent, $z \overset{\tilde{X}}{\sim} z'$ if and only if $\lambda(z,p) = \lambda'(z',p)$ for all $p \in \tilde{X}^*$.

If $\tilde{X} = X$, the $\tilde{X}$-equivalence coincides with the ordinary equivalence. An automaton $\mathfrak{A}' = (X',Y',Z',z_0',\delta',\lambda')$ is called an (α,β,γ)-homomorphism of the automaton $\mathfrak{A} = (X,Y,Z,z_0,\delta,\lambda)$ if there exist three onto-mappings

$$\alpha : X \to X'$$
$$\beta : Y \to Y'$$
$$\gamma : Z \to Z'$$

such that
$$\begin{aligned} \gamma(\delta(z,x)) &= \delta'(\gamma(z), \alpha(x)) \\ \beta(\lambda(z,x)) &= \lambda'(\gamma(z), \alpha(x)) \end{aligned} \tag{12}$$

for all $x \in X$, $z \in Z$.

If α,β,γ are one-to-one and onto, i. e. if α,β,γ are invertible, $\mathfrak{A}'$

is called an isomorphism of $\mathfrak{A}$.

If a 2-ary operation o or a 2-ary function $F: X \times X \to X$ is defined on X, we extend the operation or function to words $p, p' \in X^*$ with $|p| = |p'|$. ($|p| = n$ if $p = x_1 \dots x_n$, $|p|$ is called the length of p)

Let $p, p' \in X^*$ with $|p| = |p'|$, $p = x_1 \dots x_n$, $p' = x_1' \dots x_n'$ and let o be a 2-ary operation on X. Then we define

$$p \circ p = x_1 \circ x_1' \dots x_n \circ x_n' \tag{13}$$

If $F: X \times X \to X$ is a 2-ary function, we define

$$p \; F \; p' = F(x_1, x_1') \dots F(x_n, x_n').$$

Introductions to automata theory are e. g. /8, 11, 26/.

2. Linear automata

In this chapter we discuss basic properties of linear automata over finite and infinite fields K. The first part deals with the general response formula, the definitions of impulse response and transfer function (or matrices) of non time-dependent automata. The impulse response (or the transfer function) determines the input-output-behaviour of a non-time-dependent linear automaton with initial state 0, and by a known impulse response it is possible to construct an explicitly state dependent model of a linear automaton with the desired input-output behaviour.

It is shown that every linear automaton with an n-dimensional state vector has a finite input-output memory of at most n, $n > 0$.

The realization problem of linear automata by linear sequential circuits consisting of a few primitive components is considered. Most of the properties of linear automata mainly rely on the superposition principle, which is in force for this class of automata.

Therefore, in the second part of this chapter superposition properties of linear automata are investigated in detail. It can be shown that every time-dependent linear automaton is superponable with respect to some linear functions (e. g. with respect to $F_a(x,x') = ax + (1 - a)\,x'$, $a \in K$).

The validity of this superposition principle with respect to linear functions enables us to describe the input-output behaviour of time-dependent linear automata, even with non zero-initial state, by a set of impulse responses. Similar as in the case of non-time dependent automata it is possible to identify and to model linear time-dependent ones by a set of impulse responses.

2.1. Basic properties

A linear automaton L over a field is usually defined by the next state function

$$z(t+1) = A(t) \cdot z(t) + B(t) \cdot x(t) \tag{1}$$

and the output function

$$y(t) = C(t) \cdot z(t) + D(t) \cdot x(t). \tag{2}$$

x(t), y(t), z(t) are the input-, output- and state vectors, respectively, at time t, A(t), B(t), C(t), D(t) are time-dependent matrices with elements in K, and t is the discrete time parameter.
If none of the matrices A,B,C,D depends on t, L is called non-time-dependent and time-dependent otherwise.

l,m,n are the dimensions of the input-, output- and state vector, respectively, l,m,n > 0. z(0) is called the initial state. In the following we suppose l,m,n to be finite.
If we interpret the operation + as an arbitrary Abelian (i. e. commutative) group operation, A · z, B · x, C · z, D · x as the corresponding homomorphisms, so-called additive automata are defined by Eqs. (1) and (2) (s. /4, 37/). Most of the statements that are true for linear automata can be reformulated in the terminology of groups and additive automata.
A black-box representation of L with l input terminals and m output terminals is shown in Fig. 1.

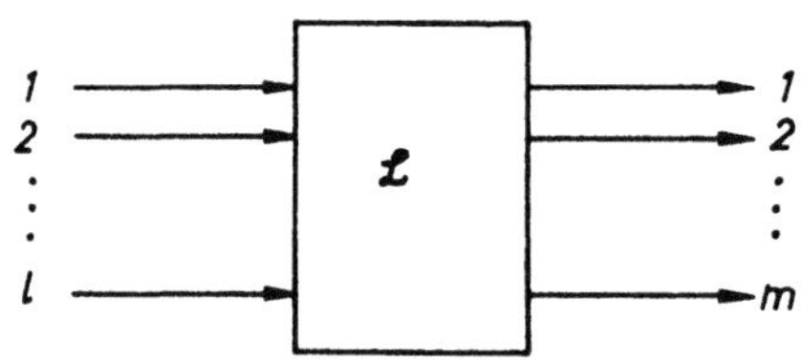

Fig.1

Each component of the input vector is associated with an input terminal and each component of the output vector is associated with an output terminal. The j-th input terminal and the k-th output terminal are sometimes referred to as the j-th input and the k-th output, respectively.

By setting

$$
\begin{aligned}
y_t &= y(t-1),\ x_t = x(t-1) \\
a_o^t &= C(t-1) \,.\, A(t-2) \dots A(0) \,.\, z(0) \\
a_1^t &= C(t-1) \,.\, A(t-2) \dots A(1) \,.\, B(0) \\
&\vdots \\
a_{t-2}^t &= C(t-1)\, A(t-2)\, B(t-3) \\
a_{t-1}^t &= C(t-1)\, B(t-2) \\
a_t^t &= D(t-1) \quad \text{for} \quad t \geq 1
\end{aligned}
\tag{3}
$$

one gets from Eqs. (1) and (2)

$$y_t = a_o^t + \sum_{i=1}^{t} a_i^t\, x_i = g_t^L(x_1,\dots,x_t). \tag{4}$$

a_o^t is an m-dimensional vector, a_i^t, $i = 1,\dots,t$ is an $m \times 1$ matrix.

If L is non-time-dependent, the a_i^t are determined by

$$
\begin{aligned}
a_o^t &= C \,.\, A^{t-1}\, z(0) \\
a_i^t &= C \,.\, A^{t-i-1}\, B, \quad i = 1,\dots,t-1 \\
a_t^t &= D
\end{aligned}
\tag{5}
$$

and in this case instead of Eq. (4) one usually writes (cf. /9, 11/)

$$y(t) = CA^t\, z(0) + \sum_{i=0}^{t} H(t-i)\, x(i) \tag{6}$$

where

$$H(t) = \begin{cases} D & \text{for } t = 0 \\ CA^{t-1}\, B & \text{for } t > 0 \end{cases} \tag{7}$$

Eqs. (6) or (4) are called the general response formula of $\mathcal{L}$.

If $z(0) = 0$, then Eq. (6) reduces to

$$y(t) = \sum_{i=0}^{t} H(t - i)\, x(i). \qquad (8)$$

y(t) is determined by the discrete convolution of H and x.

Linear automata with z(0) = 0 are referred to as quiescent linear automata.

First let us restrict ourselves to quiescent linear automata with one input terminal and one output terminal only (l = m = 1).

The output sequence that a quiescent linear automaton $\mathcal{L}_0$ produces in response to the discrete unit impulse defined by

$$u(t) = \begin{cases} 1 & \text{for } t = 0 \\ 0 & \text{for } t \neq 0 \end{cases}$$

is called its impulse response. Substituting u for x in (8) it follows that H(t) is exactly the impulse response of $\mathcal{L}_0$. On the other hand the knowledge of the impulse response H(t) is sufficient to determine the output sequence that results from the application of an arbitrary input sequence by Eq. (8). Thus the input-output behaviour of a quiescent non-time-dependent linear automaton $\mathcal{L}_0$ is uniquely determined by its impulse response H(t).

Eq. (8) can be simplified by means of the so-called $\mathcal{D}$-transform, which is introduced now. The $\mathcal{D}$-transform of a sequence $\{f(t)\}$ = f(0), f(1), f(2),... is defined by

$$\mathcal{D}(d) = \mathcal{D}[f(t)] = \sum_{j\geq 0} f(j)\, d^j$$
$$f(t) = \mathcal{D}^{-1}(d) \qquad (9)$$

$\mathcal{D}^{-1}$ denotes the inverse $\mathcal{D}$-transform. The $\mathcal{D}$-transform of an infinite sequence is a formal power series in d. A finite sequence is associated with a polynomial of finite degree. The $\mathcal{D}$-transform of an infinite periodic sequence can be computed by

$$\mathcal{D}[a_0, a_1, \ldots, a_{\mathcal{T}-1}, \underbrace{a_{\mathcal{T}}, \ldots, a_{\mathcal{T}+T-1}}_{\text{periodic}}] =$$

$$= \sum_{j=0}^{\mathcal{T}-1} a_j d^j + \frac{d^{\mathcal{T}}}{1 - d^T} \cdot \sum_{k=0}^{T-1} a_{\mathcal{T}+k}\, d^k \qquad (10)$$

and after some simple transformations it can be represented as

$$\mathcal{D}[a_0, a_1, \ldots, a_{\mathcal{T}-1}, \underbrace{a_{\mathcal{T}}, \ldots, a_{\mathcal{T}+T-1}}_{\text{periodic}}] = \frac{P(d)}{Q(d)} \tag{11}$$

where P(d) and Q(d) are polynomials of finite degree.

Since the $\mathcal{D}$-transform of a discrete convolution of two sequences is equal to the product of the $\mathcal{D}$-transforms of these sequences, Eq. (8) becomes

$$\mathcal{D}[y(t)] = I(d) \cdot \mathcal{D}[x(t)] \tag{12}$$

with

$$I(d) = \mathcal{D}[H(t)]. \tag{13}$$

I(d), the $\mathcal{D}$-transform of the impulse response, is called the transfer function of $\mathcal{L}$. The transfer function can be represented in a normal form.

Theorem 1. Let $\mathcal{L}$ be a linear automaton with one-dimensional input, one-dimensional output- and n-dimensional state vectors, then I(d) can be expressed as

$$I(d) = \frac{\beta_0 + \beta_1 d + \ldots + \beta_n d^n}{1 + \alpha_1 d + \ldots + \alpha_n d^n} \tag{14}$$

with $\beta_0, \beta_1, \ldots, \beta_n, \alpha_1, \ldots, \alpha_n \in K$.

The proof is in /11/.
There it is shown how to compute $\mathcal{D}(d)$ directly by means of the next state equation and output equation (1) and (2) for non-time-dependent automata taking into account z(0) = 0.

We consider now automata with l-dimensional input- and m-dimensional output vectors. Then we have instead of a single impulse response an l × m matrix of impulse responses $H(t) = (H_{ij}(t))$, i = 1,...,m, j = 1,...,l.
Let the input sequence be

$$x(0) = \left.\begin{matrix} 0 \\ \vdots \\ 0 \\ 1 \end{matrix}\right\} j \atop \begin{matrix} 0 \\ \vdots \\ 0 \end{matrix} \qquad x(t) = 0 \quad \text{for} \quad t \geq 1$$

Then $H_{ij}(t)$ is just the output of the i-th output terminal at time t. The transfer function I(d) in Eq. (11) has to be replaced by the transfer matrix $I(d) = (I_{ij}(d))$ with

$$I_{ij}(d) = \mathcal{D}\,[H_{ij}(t)]$$

and the $\mathcal{D}$-transform of the vector of the input and output sequences is componentwise. Thus the input-output behaviour of a quiescent linear automaton $\mathcal{L}_0$ with t-dimensional input and m-dimensional output can be described by

$$\mathcal{D}\,[y(t)] = I(d)\ .\ \mathcal{D}\,[x(t)]. \tag{15}$$

It is easy to prove that every linear automaton has a finite input output memory /9/.

<u>Theorem 2.</u> Let $\mathcal{L}$ be a linear automaton defined by Eqs. (1) and (2) (with n-dimensional state vector). Then the input-output memory of $\mathcal{L}$ is finite and equals at most r, $r \leq n$, where r is the degree of the minimum polynomial of A.

Proof. Let the minimum polynomial of A be

$$m_A(x) = \alpha_0 + \alpha_1 x + \dots + \alpha_{r-1} x^{r-1} + x^r.$$

where $r < n$.

Then, repeated applications of Eq. (6) imply

$$y(r) = CA^r \,.\, z(0) + \sum_{i=0}^{r} H(r - i)\, x(i)$$

$$= -C \sum_{j=0}^{r-1} \alpha_j A^j\, z(0) + \sum_{i=0}^{r} H(r - i)\, x(i)$$

$$= - \sum_{j=0}^{r-1} \alpha_j \{y(j) - \sum_{k=0}^{j} H(j - k)\, x(k)\} + \sum_{i=0}^{r} H(r - i)\, x(i)$$

$$= f(x(r),\dots,x(0),y(r - 1),\dots,y(0)). \blacksquare$$

Without proof we mention

<u>Theorem 3.</u> Let $\mathcal{L}$ be a linear automaton. Then there exists a minimal automaton $\mathcal{L}_{min}$ equivalent to $\mathcal{L}$ and linear.

For the procedure to determine $\mathcal{L}_{min}$ from a given $\mathcal{L}$ see /9, 10/.

Linear automata can be easily realized as linear sequential circuits. The primitive components of a linear sequential circuit over K are adders, scalers and delay-elements. (We use the terminology of /9/ here.)

An adder has l input terminals and one output terminal.
If the inputs are $x_1(t),\ldots,x_l(t)$, then the output is $x_1(t) + \ldots + x_l(t)$, where the addition is in K. If K = GF(p), the addition is modulo p.
A scaler c with a constant c, $c \in K$, has one input- and one output terminal. If the input is x(t), the output is $c \cdot x(t)$, where the multiplication is in K. c = 0 means an open connection and c = 1 means a closed connection. If K = GF(p), the multiplication is modulo p.

A delay element has one input- and one output terminal. If the input is x(t), the output is x(t - 1). The content of a delay element x(t - 1) is called its state.

Fig. 2 shows the symbols of an adder, a scaler and a delay element.

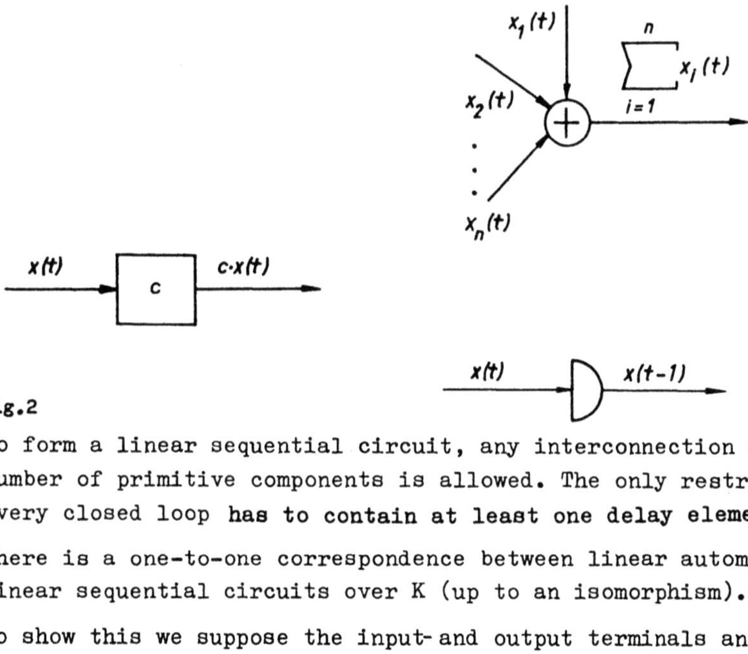

Fig.2

To form a linear sequential circuit, any interconnection of a finite number of primitive components is allowed. The only restriction is that every closed loop **has to contain at least one delay element.**

There is a one-to-one correspondence between linear automata over K and linear sequential circuits over K (up to an isomorphism).

To show this we suppose the input- and output terminals and the delay elements of the circuit under consideration to be numbered from 1 to l, 1 to m and 1 to n, respectively.

The dimensions of the input-, output- and state vector of the corresponding linear automaton are l, m and n, respectively.

The matrices $A = (A_{ij})$; $B = (B_{i,k})$, $C = (C_{r,j})$ and $D = (D_{r,k})$ reflect the connections of delay elements, the connections of input terminals and delay elements, the connections of delay elements and output terminals and the connections of input- and output terminals ($i,j = 1,\dots,n$; $k = 1,\dots,l$; $r = 1,\dots,m$). The delay element j is connected with the delay element i by a scaler A_{ij} (from j to i), thereby $A_{ij} = 0$ means no connection.

The input terminal k is connected with the delay element i by a scaler B_{ik}. The delay element j is connected with the output terminal r by a scaler C_{rj}, and the input terminal k is connected with the output terminal r by a scaler D_{rk}. Different connections leading to the same delay element or output terminal are to be summed up by an adder.
We say that a linear automaton $\mathcal{L}$ can be realized by a linear sequential circuit LSC and that an LSC can be described by a linear automaton $\mathcal{L}$.

Since the numeration of the delay elements and the input- and output terminals is arbitrary, the correspondence between linear automata and linear sequential circuits is only up to isomorphisms.

Let us consider the linear sequential circuit LSC_J represented in Fig. 3.

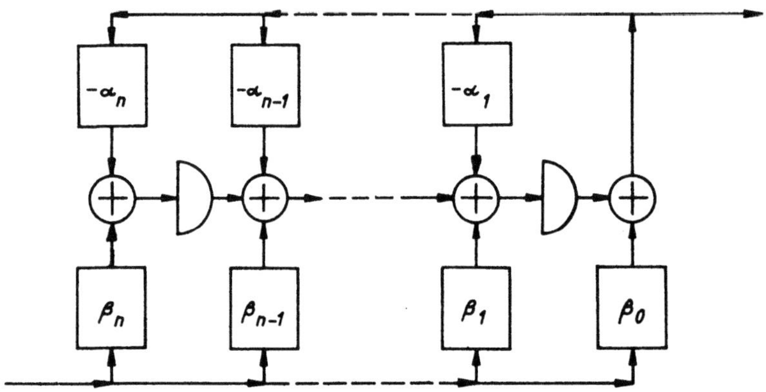

Fig.3

The LSC_J has 1 input and 1 output terminal and n delay-elements. The corresponding linear automaton $\mathcal{L}_J$ is

$$z(t+1) = A\,z(t) + B\,x(t)$$
$$y(t) = C\,z(t) + D\,x(t)$$

with

$$A = \begin{bmatrix} -\alpha_1 & 1\ 0 \dots 0\ 0 \\ -\alpha_2 & 0\ 1 \dots 0\ 0 \\ \cdot & \\ \cdot & \\ \cdot & \\ -\alpha_{n-1} & 0\ 0 \dots 0\ 1 \\ -\alpha_n & 0\ 0 \dots 0\ 0 \end{bmatrix} \qquad B = \begin{bmatrix} \beta_1 - \beta_o \alpha_1 \\ \beta_2 - \beta_o \alpha_2 \\ \cdot \\ \cdot \\ \cdot \\ \beta_n - \beta_o \alpha_n \end{bmatrix} \tag{16}$$

$C = (1\ 0 \dots 0) \quad D = (\beta_o).$

In the following we suppose the states of all delay elements to be 0 at time $t = 0$.
From Fig. 3 it is easy to see that

$$y(t) = \sum_{i=0}^{n} \beta_i\, x(t - i) - \sum_{j=1}^{n} \alpha_j\, y(t - j) \tag{17}$$

and the input-output-memory of $\mathcal{L}_{\mathcal{J}}$ equals at most n.

By the use of Eq. (7) and Definition 9 it can be shown that the transfer function $\mathcal{J}(d)$ of $\mathcal{L}_{\mathcal{J}}$ is exactly

$$\mathcal{J}(d) = \frac{\beta_o + \beta_1 d + \dots + \beta_n d^n}{1 + \alpha_1 d + \dots + \alpha_n d^n} . \tag{14}$$

Since the coefficients α_i, $i = 1,\dots,n$ and β_j, $j = 0,\dots,n$ are directly reflected in Fig. 3, the circuit shown in Fig. 3 is sometimes called a realization $LSC_{\mathcal{J}}$ of the transfer function $\mathcal{J}$ defined by Eq. (14) in its normal form.

Thus, if we know the impulse response $H(t)$ of a linear automaton $\mathcal{L}$ (maybe by an experiment), we can compute the transfer function $\mathcal{J}(d) = \mathcal{D}[H(t)]$ in its normal form (14). The normal form of the transfer function allows us explicitly to construct an $LSC_{\mathcal{J}}$ and by Eq. (16) a linear automaton $\mathcal{L}_{\mathcal{J}}$ with the desired transfer function $\mathcal{J}$. If $z(0) = 0$ for both $\mathcal{L}$ and $\mathcal{L}_{\mathcal{J}}$, the input-output-behaviour of $\mathcal{L}$ and $\mathcal{L}_{\mathcal{J}}$ coincide, i. e. $\mathcal{L}$ and $\mathcal{L}_{\mathcal{J}}$ are 0-equivalent.

Another possibility to form a linear sequential circuit LSC_H with a given periodic impulse-response $H(t) = a_o \dots a_k \underbrace{b_1 \dots b_M}_{\text{periodic}}$ is presented in Fig. 4. This realization has one closed loop only. In general the number of delay elements to realize LSC_H is larger than the number of delay-elements necessary for the realization of $LSC_{\mathcal{J}}$. The states of all delay elements are again supposed to be 0 at time t = 0.

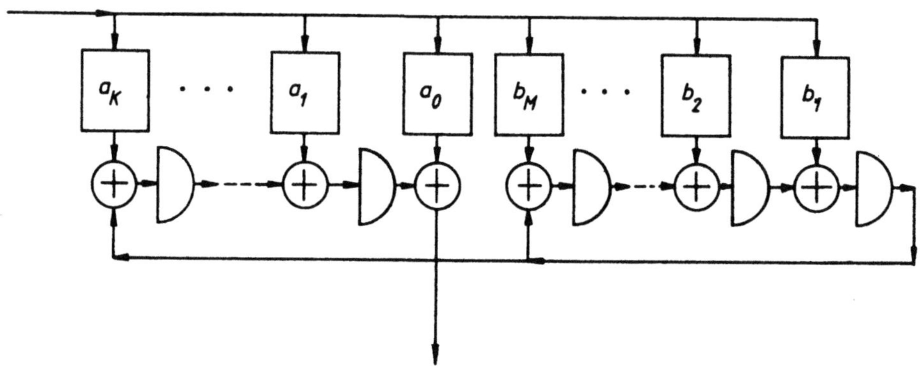

Fig.4

The following example illustrates the above results. Let $\mathcal{L}_1$ be a linear automaton over GF(3) defined by

$$\begin{aligned} z(t+1) &= \begin{bmatrix} 0 & 2 \\ 1 & 1 \end{bmatrix} \cdot z(t) + \begin{bmatrix} 2 \\ 1 \end{bmatrix} \cdot x(t) \\ y(t) &= \begin{bmatrix} 1 & 0 \\ 1 & 2 \end{bmatrix} \cdot z(t) + \begin{bmatrix} 1 \\ 0 \end{bmatrix} \cdot x(t) \end{aligned} \qquad (18)$$

The corresponding linear sequential circuit LSC_1 is shown in Fig. 5.

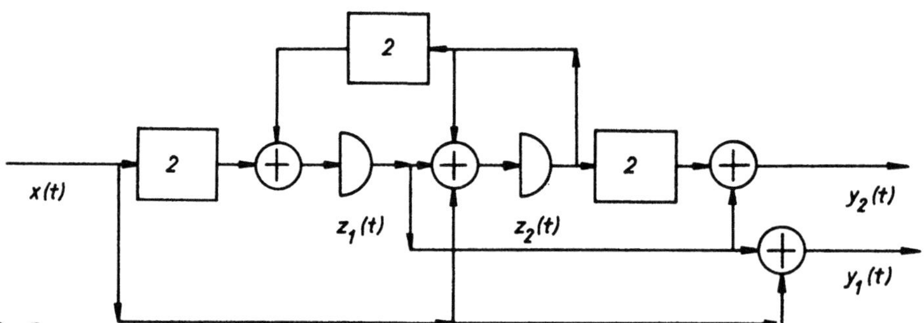

Fig.5

Since the state vector of L_1 is 2-dimensional, 2 delay elements are needed to form LSC_1. Scalers with constants 0 and 1 are drawn as open connections and closed connections, respectively.

The sequence of matrices of impulse responses is

$$H(t) = \begin{pmatrix}1\\0\end{pmatrix} \underbrace{\begin{pmatrix}2\\1\end{pmatrix} \begin{pmatrix}2\\2\end{pmatrix} \begin{pmatrix}0\\1\end{pmatrix} \begin{pmatrix}1\\2\end{pmatrix} \begin{pmatrix}1\\1\end{pmatrix} \begin{pmatrix}0\\2\end{pmatrix}}_{\text{periodic}} \tag{19}$$

This sequence can be computed by Eq. (7) and it is the response of $\mathcal{L}_1$ to the input sequence 1 0 0 0 ...

$I_{11}(d)$ is determined by

$$\mathfrak{I}_{11}(d) = \mathcal{D}[H_{11}(t)] = \mathcal{D}[1 \underbrace{2\ 2\ 0\ 1\ 1\ 0}_{\text{periodic}}]$$

$$= 1 + d\left[\frac{2 + 2d + d^3 + d^4}{1 - d^6}\right] = \frac{1 + d + d^2}{1 + 2d + d^2}\,. \tag{20}$$

$\mathfrak{I}_{21}(d)$ is determined by

$$\mathfrak{I}_{21}(d) = \mathcal{D}[H_{21}(t)] = \mathcal{D}[0 \underbrace{1\ 2}_{\text{periodic}}] = d\left(\frac{1 + 2d}{1 - d^2}\right) = \frac{d}{1 + d} \tag{21}$$

and finally we have

$$\mathfrak{I}(d) = \begin{pmatrix} \dfrac{1 + d + d^2}{1 + 2d + d^2} \\ \\ d + d^2 \end{pmatrix}. \tag{22}$$

The realizations of $\mathfrak{I}_{11}(d)$ and $\mathfrak{I}_{21}(d)$ are represented in Fig. 6a and 6b (In GF(3) we have $-1 = 2$, $-2 = 1$).

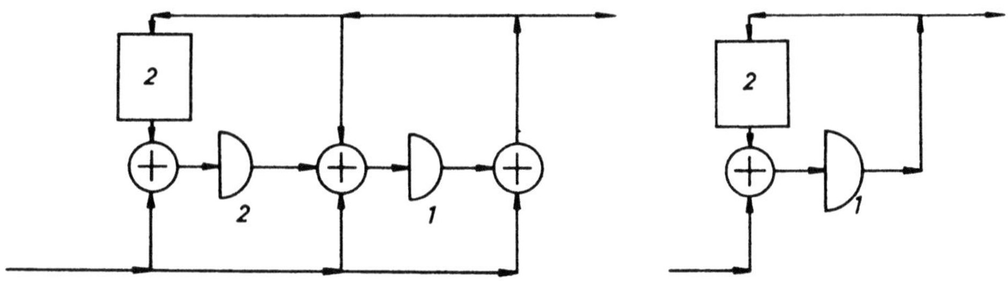

Fig.6

If we connect the input terminals of Fig. 6a and 6b, then we get a realization $LSC_{\mathfrak{I}}$ of $\mathfrak{I}$ as shown in Fig. 7.

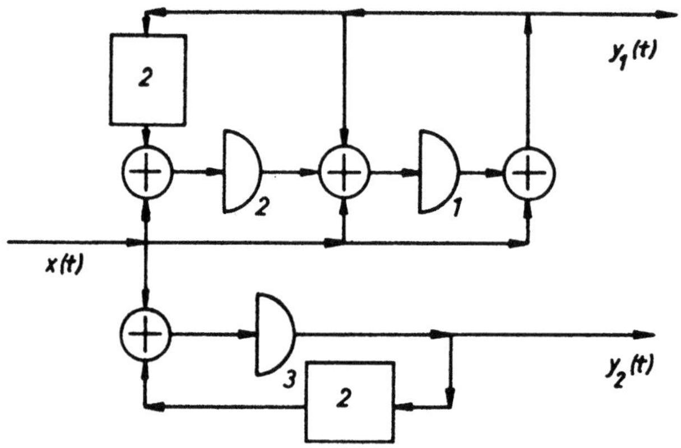

Fig.7

Of course the $LSC_{\mathfrak{J}}$ represented in Fig. 7 is not minimal. The corresponding automaton $\mathfrak{L}_{\mathfrak{J}}$ is

$$
\begin{aligned}
z(t+1) &= \begin{pmatrix} 1 & 1 & 0 \\ 2 & 0 & 0 \\ 0 & 0 & 1 \end{pmatrix} \cdot z(t) + \begin{pmatrix} 2 \\ 0 \\ 1 \end{pmatrix} \cdot x(t) \\
y(t) &= \begin{pmatrix} 1 & 0 & 0 \\ 0 & 0 & 1 \end{pmatrix} \cdot z(t) + \begin{pmatrix} 1 \\ 0 \end{pmatrix} \cdot x(t) .
\end{aligned} \tag{23}
$$

The realizations $LSC_{H_{11}}$ and $LSC_{H_{21}}$ are shown in Fig. 8a and 8b.

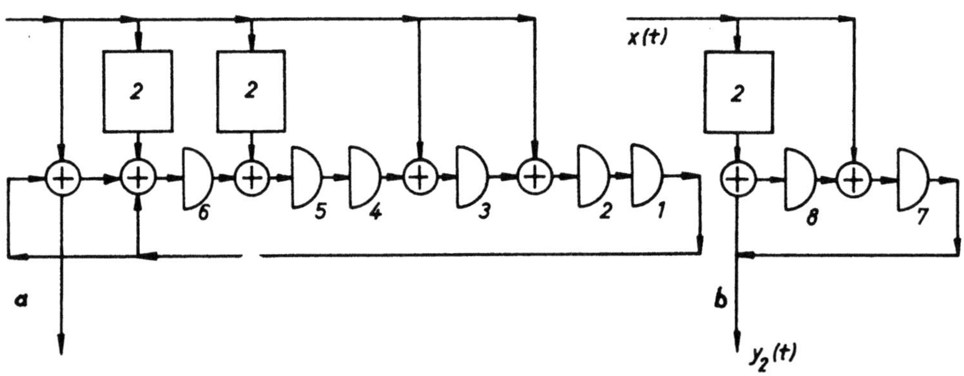

Fig.8

By connecting the input terminals of Figs. 7a and 7b we get the realization LSC_H, and the corresponding automaton $\mathcal{L}_H$ can be described by

$$z(t+1) = \begin{bmatrix} 0&1&0&0&0&0&0&0 \\ 0&0&1&0&0&0&0&0 \\ 0&0&0&1&0&0&0&0 \\ 0&0&0&0&1&0&0&0 \\ 0&0&0&0&0&1&0&0 \\ 1&0&0&0&0&0&0&0 \\ 0&0&0&0&0&0&0&1 \\ 0&0&0&0&0&0&1&0 \end{bmatrix} \cdot z(t) + \begin{bmatrix} 0 \\ 1 \\ 1 \\ 0 \\ 2 \\ 2 \\ 1 \\ 2 \end{bmatrix} \cdot x(t)$$

$$y(t) = \begin{pmatrix} 1&0&0&0&0&0&0&0 \\ 0&0&0&0&0&0&1&0 \end{pmatrix} \cdot z(t) + \begin{pmatrix} 1 \\ 0 \end{pmatrix} \cdot x(t) \tag{24}$$

The impulse responses H(t) are identical for $\mathcal{L}_1$, $\mathcal{L}_J$ and $\mathcal{L}_H$.
Now we apply the $\mathcal{D}$-transform to compute the output sequence for a given input sequence.

Let $[x(t)] = \underbrace{1\ 2\ 1}_{\text{periodic}}$. Then we have by use of Eq. (10)

$$\mathcal{D}[x(t)] = X(d) = \frac{1 + 2d + d^2}{1 - d^3}$$

and by Eqs. (15) and (22)

$$\begin{bmatrix} y_1(d) \\ y_2(d) \end{bmatrix} = \begin{bmatrix} \dfrac{1 + d + d^2}{1 + 2d + d^2} \\ d + d^2 \end{bmatrix} \cdot \frac{1 + 2d + d^2}{1 - d^3} = \begin{bmatrix} \dfrac{1}{1 - d} \\ d\ \dfrac{1 + d^3}{1 - d^3} \end{bmatrix}, \tag{25}$$

which implies

$$y_1(t) = \underbrace{1}_{\text{periodic}} \qquad y_2(t) = 0\ \underbrace{1\ 0\ 0\ 1}_{\text{periodic}}.$$

2.2. Linear superposition of linear automata

The second part of this chapter deals with superposition properties of linear automata (s. /18/).
First let us consider two different input sequences $\{x(t)\} = x(0)x(1)x(2) \ldots$ and $\{x'(t)\} = x'(0)x'(1)x'(2)$ of a quiescent linear automaton

$\mathcal{L}$. The corresponding output sequences $\{y(t)\} = y(0)\ y(1)\ y(2) \ldots$ and $\{y'(t)\} = y'(0)\ y'(1)\ y'(2) \ldots$ that $\mathcal{L}$ produces in response to the input sequences $\{x(t)\}$ and $\{x'(t)\}$ are determined by Eq. (8).

Since Eq. (8) is linear, the output sequences that result from the application of the sum of the input sequences
$\{x(t) + x'(t)\} = (x(0) + x'(0))(x(1) + x'(1))(x(2) + x'(2)) \ldots$ are the sum of the output sequences $\{y(t) + y'(t)\} = (y(0) + y'(0))(y(1) + y'(1))(y(2) + y'(2)) \ldots$

This fact can be interpreted that a superposition principle with respect to the operation "addition in K" is in force.

If x and y are l- and m-dimensional vectors, $l,m > 1$, the operation addition is to be understood as the componentwise addition. In the following let $l = m = 1$ for simplicity.

If $\mathcal{L}$ is a quiescent linear automaton, then the two circuits represented in fig. 9 are obviously equivalent.

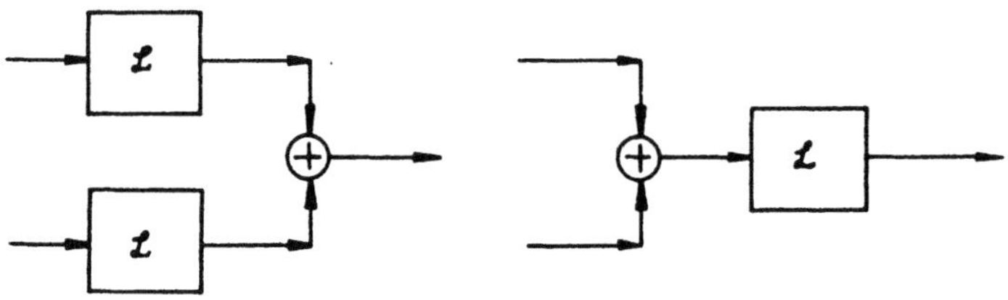

Fig.9

It is easy to see that the operation addition is not the only one for which a superposition principle is valid.

Thus the output sequence that $\mathcal{L}$ produces in response to the input sequence $\{a \cdot x(t) + bx'(t)\}$ is $\{a \cdot y(t) + by'(t)\}$, $a,b \in K$.
If we denote $(a \cdot x + bx') = F_{a,b}(x,x')$, the circuits represented in Fig. 10 are equivalent, and $\mathcal{L}$ is superponable with respect to $F_{a,b}$.

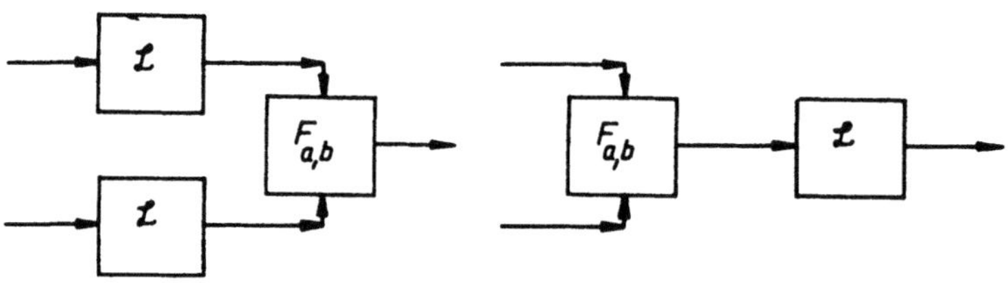

Fig.10

The following theorems deal with the question with respect to which functions F linear automata are superponable. The automata under consideration do not have to be quiescent or time independent.

Theorem 4. Let $\mathcal{L}$ be a linear automaton over GF(p) with one input- and one output terminal, let $\mathcal{L}$ be essential (c. f. ch. 1) and F an essential p-valued logical function of two variables. Then $\mathcal{L}$ is F-superponable if and only if F has the form

$$F(x,x') = ax + bx' + c, \qquad a,b,c \in GF(p), \quad a \neq 0, \quad b \neq 0 \tag{26}$$

and the condition

$$(a + b - 1)\, a_o^t = c(-1 + \sum_{i=1}^{t} a_i^t) \tag{27}$$

holds for all $t > 0$.

Proof. It is easy to check that every function F defined by Eq. (26) and taking into account Eq. (27) satisfies

$$F(a_o^t + \sum_{i=1}^{t} a_i^t\, x_i,\ a_o^t + \sum_{i=1}^{t} a_i x_i') = a_o^t + \sum_{i=1}^{t} a_i^t\, F(x_i, x_i'). \tag{28}$$

On the other hand, since F is a function over GF(p), it can be uniquely expressed as

$$F(x,x') = \sum_{i,j=1}^{p-1} \alpha_{ij}\, x^i x^j \tag{29}$$

with $\alpha_{ij} \in GF(p)$ (cf. /38/).

As $\mathcal{L}$ is essential, there has to exist a time t such that at least two different coefficients a_j^t, a_k^t in Eq. (4) do not equal zero. Substituting the representation (29) in Eq. (28) and comparing the coefficients, we get the result.

A similar theorem can be proved if K is the infinite field of real numbers.

Theorem 4'. Let $\mathcal{L}$ be an essential linear automaton over the infinite field K of the real numbers and let F be an essential differentiable real function of two variables. Then $\mathcal{L}$ is F-superponable if and only if F has the form

$$F(x,x') = ax + bx' + c \qquad a,b,c \in K, \quad a \neq 0, \quad b \neq 0 \tag{26'}$$

and the condition

$$(a + b - 1)\, a_o^t = c(-1 + \sum_{i=1}^{t} a_i^t)$$

holds for all $t > 0$.

Proof. The first part of the proof is identical with the first part of the proof of Theorem 4.
Let

$$F(a_o^t + \sum_{i=1}^{t} a_i^t x_i,\ a_o^t + \sum_{i=1}^{t} a_i^t x_i') = a_o + \sum_{i=1}^{t} a_i^t F(x_i, x_i). \tag{28}$$

Since $\mathcal{L}$ is essential, there have to be for at least one time t coefficients a_j^t, a_k^t with $a_j^t \neq 0$, $a_k^t \neq 0$. By setting $x_i = x_i' = 0$ for $i \neq j,k$ Eq. (28) reduces to

$$F(a_o^t + a_j^t x_j + a_k^t x_k,\ a_o^t + a_j^t x_j' + a_k^t x_k') = a_o^t + a_j^t F(x_j, x_j') + + a_k^t F(x_k, x_k'). \tag{29}$$

Differentiating with respect to x_j Eq. (29) becomes

$$F_1(a_o^t + a_j^t x_j + a_k^t x_k,\ a_o^t + a_j^t x_j' + a_k^t x_k') = F_1(x_j, x_j') \tag{30}$$

and with $x_j = x_j' = -\frac{a_o^t}{a_j^t}$, $a_k^t x_k = x$, $a_k^t x_k = x'$

$$F_1(x,x') = F_1(-\frac{a_o^t}{a_j^t} - \frac{a_o^t}{a_j^t}) = a = \text{const.} \tag{31}$$

Differentiating Eq. (29) with respect to x_j' we obtain in a similar way

$$F_2(x,x') = F_2(-\frac{a_o^t}{a_j^t},\ -\frac{a_o^t}{a_j^t}) = a = \text{const,} \tag{32}$$

which implies

$F(x,x') = ax + by + c$ (26).

Substituting Eq. (26) in (28) and comparing the coefficients we get the condition (27).■

As a special case of Theorems 4 and 4' we have

Conclusion 5. Each linear automaton L over a field K is F_a superponable with

$$F_a(x,x') = ax + (1 - a)\, x' \qquad a \in K \tag{33}$$

If $a_o^t = 0$ for all $t > 0$ (i. e. $z(0) \sim 0$), then $\mathcal{L}$ is $F_{a,b}$ superponable with

$$F_{a,b}(x,x') = ax + bx'. \tag{34}$$

We have to mention that for GF(2) neither $F_0(x,x') = x'$ nor $F_1(x,x') = x$ are essential 2-ary functions, and the case GF(2) requires special considerations. The only two linear essential functions of two variables in GF(2) are $F(x,x') = x + x'$ (antivalence) and $\tilde{F}(x,x') = 1 + x + x'$ (equivalence).

It is easy to see that the automaton $\mathcal{L}_2$ with initial state $z(0) = \binom{1}{0}$ defined by

$$\begin{aligned} z(t+1) &= \begin{pmatrix} 0 & 1 \\ 1 & 0 \end{pmatrix} \cdot z(t) + \begin{pmatrix} 1 \\ 1 \end{pmatrix} \cdot x(t) \\ y(t) &= (10) \cdot z(t) \end{aligned} \tag{35}$$

is neither F- nor $\tilde{F}$-superponable.

In the case of GF(2) we have

Conclusion 6. A linear automaton $\mathcal{L}$ over GF(2) is F- or $\tilde{F}$-superponable with $F(x,x') = x + x'$, $\tilde{F}(x,x') = 1 + x + x'$ if and only if for all $t > 0$ $a_o^t = 0$ (i. e. $g_t^{\mathcal{L}}(0 \ldots 0) = 0$) or

$$a_o^t = 1 + \sum_{i=1}^{t} a_i^t \quad (\text{i. e. } g_t^{\mathcal{L}}(1 \ldots 1) = 1).$$

Proof. This conclusion follows immediately from Condition (27) of Theorem 4. ∎

Linear automata over GF(2) are special binary automata. Superposition properties of binary automata are investigated in chapter 3.2.

The superposition properties of linear automata allow us to describe the input-output-behaviour of these automata by means of impulse responses or in some sense generalized impulse responses even if the automata under consideration are time dependent and not quiescent.

Thus the complete input-output-behaviour of linear automata is determined by their response to some special input sequences.

Theorem 7. Let $\mathcal{L}$ be a linear (time dependent) automaton over K, then for all $x_1,\dots,x_t \in X$ $\lambda(x_1\dots x_t)$ is determined by $\lambda(0\dots0)$, $\lambda(a_1 0\dots0),\dots,$ $\lambda(0\dots0a_t)$ with $a_i \neq 0$, $i = 1,\dots,t$.

Proof. Since $(0^{i-1}b0^{t-i}) = (0^{i-1}a_i0^{t-i})\ F_{\frac{b}{a_i}}\ (0^t)$ and since $\mathcal{L}$ is F_α-superponable for $\alpha \in X$, all output words $\lambda(0^{i-1}b0^{t-i})$ are determined by $\lambda(0^{i-1}b0^{t-1}) = \lambda(0^{i-1}a_i0^{t-1})\ F_{\frac{b}{a_i}}\ \lambda(0^t)$, $b \in X$, $i = 1,\dots,t-1$.

Let us fix some $u \in X$ with $u \neq 0$ and $u \neq 1$.
Then we have

$$(x_1\dots x_t) = (b_1^1 \dots b_{t-1}^1 0)\ F_u(0^{t-1}b_t^{1'})$$

where

$$b_j^1 = \frac{x_j}{u} \quad j = 1,\dots,t-1, \quad b_t^{1'} = \frac{x_t}{1-u},$$

$$(b_1^1\dots b_{t-1}^1 0) = (b_1^2\dots b_{t-2}^2 0^2)\ F_u(0^{t-2}b_{t-1}^{2'}0)$$

where

$$b_j^2 = \frac{b_j^1}{u} \quad j = 1,\dots,t-2, \quad b_{t-1}^{2'} = \frac{b_{t-1}^{'}}{1-u}$$

$$\vdots$$

$$(b_1^{t-2}\ b_2^{t-2}\ 0^{t-2}) = (b_1^{t-1}\ 0^{t-1})\ F_u(0b_2^{t-1'}0^{t-2})$$

where

$$b_1^{t-1} = \frac{b_1^{t-2}}{u}, \quad b_2^{t-1'} = \frac{b_2^{t-2}}{1-u},$$

and therefore $x_1\dots x_t$ can be expressed as $x_1\dots x_t =$

$$\{\dots[(b_1^{t-1}0^{t-1})\ F_u(0b_2^{t-1'}0^{t-2})]\ F_u[0^2b_3^{t-2'}0^{t-3}]\dots\}\ F_u\{0^{t-1}b_t^{1'}\}.$$

Since $\mathcal{L}$ is F_u-superponable, $\lambda(x_1\dots x_t)$ can be computed by
$\lambda(x_1\dots x_t) =$

$$\{\dots[\lambda(b_1^{t-1}0^t)\ F_u\ \lambda(0b_2^{t-1}0^{t-2})]\ F_u\ [\lambda(0^2b_3^{t-2'}0^{t-3})]\dots\}\ F_u\ \{\lambda(0^{t-1}b_t^{1'})\}.$$

■

If $\mathcal{L}$ is an F- or $\tilde{F}$-superponable automaton over GF(2) with $F(x,x') = x + x$ and $\tilde{F}(x,x') = 1 + x + x'$, respectively, then $\lambda(x_1\dots x_t)$ is determined by $\lambda(0\dots0)$, $\lambda(1\ 0\ \dots\ 0),\dots$ $\lambda(0\dots0\ 1)$ respectively

$\lambda(1\ldots1)$, $\lambda(0\ 1\ldots1),\ldots,\ \lambda(0\ldots0\ 1)$.

Such linear automata for which the input-output-behaviour can be described by a finite set of periodic inpulse responses are of special interest.

<u>1.</u> Let $\mathcal{L}$ be a non-time-dependent automaton with $z(0) \sim 0$. Since $\lambda(0^i 1\ 0^\infty) = 0^i\ \lambda(1\ 0^\infty)$, $i \geq 0$, the input-output-behaviour of $\mathcal{L}$ is determined by $\lambda(1\ 0^\infty)$. If $\mathcal{L}$ is finite, the impulse response $\lambda(1\ 0^\infty)$ is periodic. The realizations LSC_I or LSC_H (cf. Figs. 3 and 4) and the corresponding linear automata L_I or L_H can be formed as described before.

If $z(0)$ is not equivalent to 0, the input-output-behaviour of $\mathcal{L}$ is determined by $\lambda(0^\infty)$ and $\lambda(1\ 0^\infty)$. In this case we denote $\lambda^o(1\ 0^\infty) = \lambda(1\ 0^\infty) - \lambda(0^\infty)$. Then $\lambda^o(0^i 1\ 0^\infty) = 0^i\ \lambda^o(1\ 0^\infty)$.

If $\mathcal{L}$ is finite, then $\lambda(0^\infty)$ and $\lambda(1\ 0^\infty)$, and therefore $\lambda^o(1\ 0^\infty)$ are periodic, $\lambda(0^\infty) = a_1\ldots a_l\ \underbrace{b_1\ldots b_k}_{\text{periodic}}$, $\mathcal{L}$ can be realized as shown

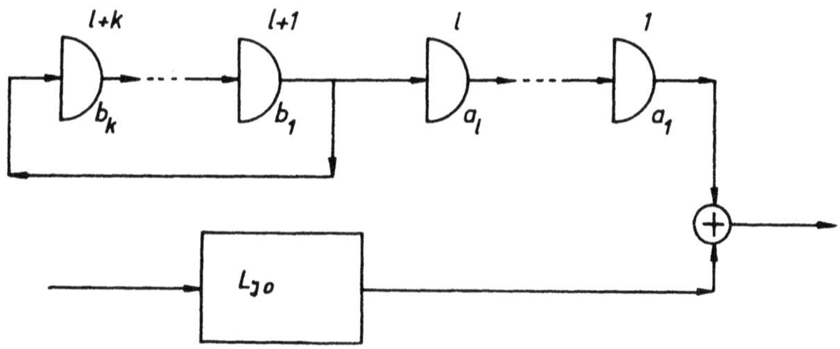

Fig.11

$\mathcal{L}_{I0}$ is a realization of $\lambda^o(1\ 0^\infty)$, the initial states of the delay elements $1,\ldots,l\ \ l+1,\ldots,l+k$ are $a_1,\ldots,a_l$ and $b_1\ldots b_k$, respectively.

<u>2.</u> Let $\mathcal{L}$ be a periodically time-dependent automaton with $z(0) \sim 0$ with a period T, $T > 0$ and

$$\lambda(0^K 1\ 0^\infty) = 0^k\ \lambda^{k(\text{mod } T)}(1\ 0^\infty).$$

The corresponding transfer functions are denoted by $\mathcal{J}_0,\ldots,\mathcal{J}_{T-1}$, respectively. The linear sequential circuit represented in Fig. 12 has the desired impulse responses and, since it is linear, the desired input-output-behaviour. At time t the switch S has to be in position t(mod T).

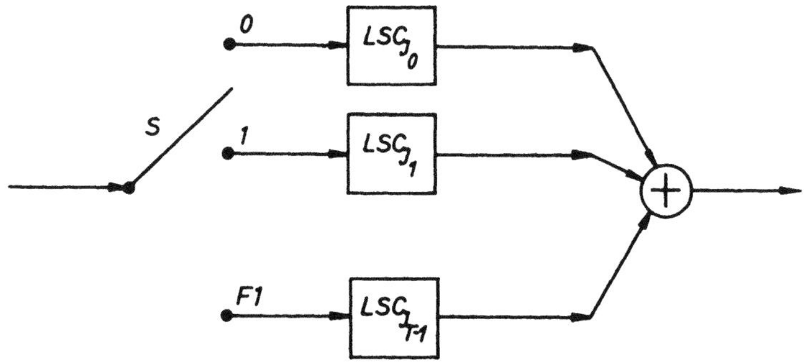

Fig.12

The condition z(0) ~ 0 is not necessary for the construction of an explicitly state-dependent model.

3. Automata superponable with respect to pairs of operations

This chapter deals with the superposition of automata with respect to a pair of operations (o,▽) or a pair of functions (F,G) which are different from the ordinary addition.

The first part is dedicated to the general formulation of the problem and to general solutions. As a main result it is proved that the next-state function and the output function of a (o,▽)-superponable automaton have to satisfy functional equations known as equations of generalized bisymmetry. Some interrelations of properties of the operations o and ▽ of (o,▽)-superponable automata are discussed.

The following parts concern special classes of superponable automata.

Binary automata are investigated in the second part. For all pairs of Boolean functions F and G the class of (binary) (F,G)-superponable automata is characterized. This result heavily relies on the fact that every Boolean function can be uniquely represented as a polynomial of finite degree over GF(2), known as Shegalkin's polynomial.

A state-dependent description of (F,G)-superponable automata is given and it is shown how to derive an explicitly state-dependent model of a (F,G)-superponable automaton by use of the so-called generalized impulse responses.

The third part deals with (o,▽)-linear automata. The operations o and ▽ are supposed to be associative and commutative. Analogous to the case of ordinary linearity the notions (o,▽)-linear functions and (o,▽)-linear automata are introduced.

It is shown that the class of (o,∇)-superponable automata and the class of (o,∇)-linear automata coincide. This is the reason why the characterization of (o,∇)-superponable automata needs only the solution of a relatively simple functional equation $h(x \circ x') = h(x) \nabla h) x')$, where h is a function only.

If o and ∇ are rational operations, this functional equation can be reduced to an ordinary differential equation of first order.

If o and ∇ are quasiadditions, i. e. if $x \circ x' = \varphi(\varphi^{-1}(x) + \varphi^{-1}(x'))$ and $y \nabla y' = \psi(\psi^{-1}(y) + \psi^{-1}(y'))$ where φ and ψ are strictly monotonic functions, a canonic representation of (o,∇)-linear automata as a serial connection of two combinational functions φ^{-1} and ψ and a linear automaton is derived. In this case different (o,∇)-linear automata differ only in their linear parts.

It is shown how the input-output-behaviour of (o,∇)-superponable automata can be described by means of generalized impulse responses.

At the end of this chapter $(o,\nabla,\square)$-linear automata, which are (o,∇)-superponable as a generalization of (o,∇)-linear automata, are introduced. Thereby the operations o, ∇ and $\square$ are supposed to satisfy a special equation of generalized bisymmetry.

3.1. Definitions and basic results

Now we introduce the superposition principle of abstract automata with respect to a pair of 2-ary operations or functions. In our further discussion it is convenient to have different formulations of the same facts. Therefore we introduce three different definitions of superponability, which of course have to be equivalent.

Definition 1. Let $\mathfrak{A} = (X,Y,Z,z_0,\delta,\lambda)$ and let o and ∇ be two 2-ary operations on X and Y, respectively. Then $\mathfrak{A}$ is called (o,∇)-superponable if for all $t > 0$, $x_1,\ldots,x_t$, $x_1',\ldots,x_t' \in X$

$$g_t(x_1 \circ x_1',\ldots,x_t \circ x_t') = g_t(x_1,\ldots,x_t) \nabla g_t(x_1',\ldots,x_t') . \qquad (1)$$

If $X = Y$, $x \circ x' = x \nabla x'$ for $x,x' \in X$, we say that $\mathfrak{A}$ is o-superponable. For given operations o and ∇ Eq. (7) is a system of functional equations for the functions g_t, $t = 1,2,\ldots$

If we use instead of the operations o and ∇ 2-ary functions F and G, then Definition 1 becomes

Definition 1'. Let $\mathfrak{A} = (X,Y,Z,z_o,\delta,\lambda)$ and let $F : X \times X \to X$, $G : Y \times Y \to Y$ be 2-ary functions. Then $\mathfrak{A}$ is called (F,G)-superponable if for all $t > 0$, $x_1,\dots,x_t$, $x'_t,\dots,x'_t \in X$

$$g_t(F(x_1,x'_1),\dots,F(x_t,x'_t)) = G(g_t(x_1,\dots,x_t),\ g_t(x'_1,\dots,x'_t)) \qquad (2)$$

If $X = Y$ and $F = G$, we say that $\mathfrak{A}$ is F-superponable.
Setting $x \circ x' = F(x,x')$ for $x,x' \in X$ and $y \triangledown y' = G(y,y')$ for $y,y' \in Y$, Eqs. (1) and (2) become identical.

Definition 1 is in terms of input symbols. It can be reformulated in terms of input words.

Definition 1". Let $\mathfrak{A}$, o, $\triangledown$ be as in Def. 1. Then $\mathfrak{A}$ is called (o,$\triangledown$)-superponable if for all $p,p' \in X$ with $|p| = |p'|$

$$\lambda(z_o,\ p \circ p') = \lambda(z_o,p) \triangledown \lambda(z_o,p') \qquad (3)$$

Eq. (1) is the componentwise notation of Eq. (3), and λ is the extended output-function of $\mathfrak{A}$.

If o and/or $\triangledown$ are time-dependent operations, we write o_t, $\triangledown_t$.

The circuits represented in Fig. 1 and Fig. 2 are mutually equivalent if and only if the automaton $\mathfrak{A}$ is (o,$\triangledown$)-superponable or (F,G)-superponable, respectively.

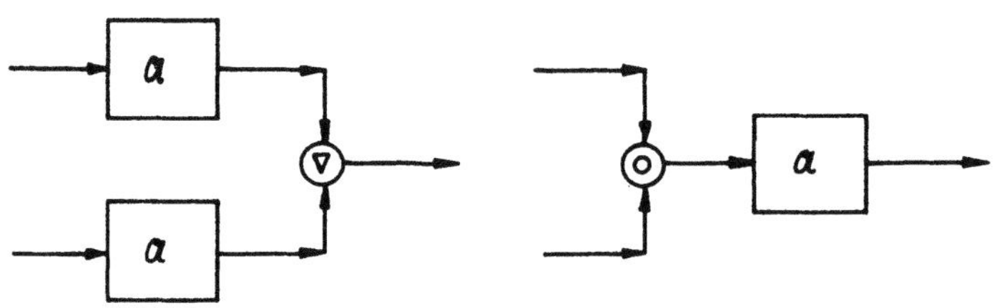

Fig.1

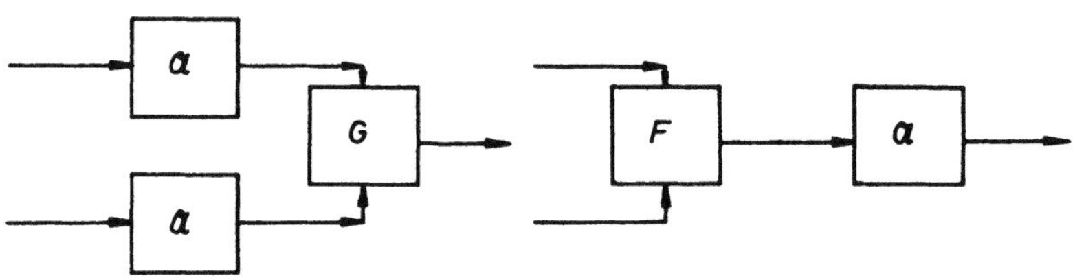

Fig.2

As it is seen from Fig. 1, (o,$\triangledown$)-superponability of automata may be used to simplify automata circuits.

(Here as in the following we do not mention that the same is true for (F,G)-superponability.)

The following theorems and conclusions deal with the question: What is the internal structure of a (o,∇)-superponable automaton? Or more precisely: **What conditions have the next state- and output functions of a** superponable automaton to satisfy?

Theorem 1. Let $\mathfrak{A} = (X,Y,Z,z_o,\delta,\lambda)$ be a reduced automaton, and let $Z_t = \{z \,/\, \exists\, p \in X^t \wedge z = \delta(z_o,p)\}$.
Then $\mathfrak{A}$ is (o,∇)-superponable if and only if there exists a 2-ary operation $\square$ on Z such that for all $t > 0, x_1,\ldots,x_t,x'_1,\ldots,x'_t \in X$, $z,z' \in Z_t$

$$1.\quad \delta(z_o,x_1 \circ x'_1 \ldots x_t \circ x'_t) = \delta(z_o x_1 \ldots x_t) \,\square\, \delta(z_o x'_1 \ldots x'_t) \qquad (4)$$

$$2.\quad \lambda(z \,\square\, z', x \circ x') = \lambda(z,x) \,\nabla\, \lambda(z',x'). \qquad (5)$$

Proof a) Let $\mathfrak{A}$ be (o,∇)-superponable.
Then the proof needs

Lemma 2. Let be 1. $\mathfrak{A} = (X,Y,Z,z_o,\delta,\lambda)$ a (o,∇)-superponable automaton,
2. $\tilde{X} = \{x/\exists\, x_1,x_2 \in X \wedge x = x_1 \circ x_2\}$
3. $z_p = \delta(z_o,p)$ for $p \in X^*$
4. $p_1,p'_1,p_2,p'_2 \in X^*$ with $|p_1| = |p'_1|$ and $|p_2| = |p'_2|$

Then $z_{p_1} \sim z_{p_2}$ and $z_{p'_1} \sim z_{p_2'}$ imply $z_{p_1 o p'_1} \overset{\tilde{X}}{\sim} z_{p_2 o p'_2}$.

Proof of Lemma 2. $\mathfrak{A}$ is (o,∇)-superponable if and only if for all $p,p' \in X^*$ with $|p| = |p'|$

$$\lambda(z_o, p \circ p') = \lambda(z_o,p) \,\nabla\, \lambda(z_o,p'). \qquad (3)$$

Let $p = r\,s$, $p' = r's'$ with $|r| = |r'|$, $|s| = |s'|$. Then from Eq. (3) we obtain

$$\lambda(\delta(z_o,r \circ r'), s \circ s') = \lambda(\delta(z_o,r), s) \,\nabla\, \lambda(\delta(z_o,r'), s'). \qquad (5)$$

Substituting $r = p_i$, $r' = p'_i$, $i = 1,2$, Eq. (5) becomes

$$\lambda(z_{p_i o p'_i}, s \circ s') = \lambda(z_{p_i},s) \,\nabla\, \lambda(z_{p'_i},s'). \qquad (6)$$

Since $z_{p_1} \sim z_{p_2}$ and $z_{p'_1} \sim z_{p'_2}$ we have $\lambda(z_{p_1},s) = \lambda(z_{p_2},s)$, $\lambda(z_{p'_1},s') = \lambda(z_{p'_2},s')$ and therefore $\lambda(z_{p_1 o p'_1},s \circ s') = \lambda(z_{p_2 o p'_2},s \circ s')$, $s \circ s' \in \tilde{X}^*$, which finishes the proof of Lemma 2.

Now we define a 2-ary operation $\square$ on Z with the property $z, z' \in Z_t$ implies $z \square z' \in Z_t$.

Let $z_t^1, z_t^2 \in Z_t$, $z_t^1 = \delta(z_o, p^1)$, $z_t^2 = \delta(z_o, p^2)$, $|p^1| = |p^2| = t$.
Then we put $z_t^1 \square z_t^2 = z_{p^1 o p^2}$.

If $z_t^1 = \delta(z_o, p^3)$, $z_t^2 = \delta(z_o, p^4)$, $|p^3| = |p^4| = t$, then we have $z_{p^3 o p^4} \sim z_{p^1 o p^2}$ by virtue of Lemma 2. Since $\mathfrak{A}$ is minimal, ,we obtain $z_{p^3 o p^4} = z_{p^1 o p^2}$.

Let now $z_{t'}^1, z_{t'}^2 \in Z_{t'}$, $z_{t'}^1 = \delta(z_o, p^{1'})$, $z_{t'}^2 = \delta(z_o, p^{2'})$,
$|p^1| = |p^2| = t' \neq t$
and $z_{t'}^1 = z_t^1$, $z_{t'}^2 = z_t^2$. Again by Lemma 2 we get $z_{t'}^1 \square z_{t'}^2 = z_{p^{1'} o p^{2'}} \sim z_{p_1 o p_2} = z_t^1 \square z_t^2$. Since $\mathfrak{A}$ is minimal, we again conclude $z_{t'}^1 \square z_{t'}^2 = z_t^1 \square z_t^2$ and the definition of $\square$ is without contradiction.
Setting $p^1 = x_1 \ldots x_t$, $p^2 = x_1' \ldots x_t'$ Eq. (4) results from the definition of $\square$.

Let $z, z' \in Z_t$ and $z = \delta(z_o, p)$, $z' = \delta(z_o, p')$ $p = x_1 \ldots x_{t-1}$, $p' = x_1' \ldots x_{t-1}'$.
Since $\mathfrak{A}$ is supposed to be (o,∇)-superponable, we have

$$\lambda(z \square z', x \circ x') = \lambda(\delta(z_o,p) \square \delta(z_o,p'), x \circ x') =$$
$$= \lambda(\delta(z_o, p \circ p'), x \circ x') = g_t(x_1 \circ x_1', \ldots, x_{t-1} \circ x_{t-1}', x \circ x') =$$
$$= g_t(x_1, \ldots, x_{t-1}, x) \nabla g_t(x_1' \ldots x_{t-1}', x')$$
$$= \lambda(z,x) \nabla \lambda(z',x').$$

b) If the conditions of Theorem 1 are satisfied, $\mathfrak{A}$ is obviously (o,∇)-superponable. ∎

From Theorem 1 we obtain

<u>Conclusion 3.</u> Let $\mathfrak{A} = (X,Y,Z,z_o,\delta,\lambda)$ be a minimal (o,∇)-superponable automaton. Then the state automaton $\tilde{\mathfrak{A}} = (X,Z,z_o,\delta)$ is $(o,\square)$-superponable, where $\square$ is defined as before.

<u>Theorem 4.</u> Let $\mathfrak{A}$, Z_t and $\square$ be as in Theorem 1.
Then $\mathfrak{A}$ is (o,∇)-superponable if and only if for all $t > 0$, $z, z' \in Z_t$, $x, x' \in X$

1. $\delta(z \square z', x \circ x') = \delta(z,x) \square \delta(z',x')$ (7)
2. $\lambda(z \square z', x \circ x') = \lambda(z,x) \nabla \lambda(z',x')$ (8)

Proof. Because of Theorem 1 **we only** have to show:

$$\delta(z \square z', x \circ x') = \delta(z,x) \square \delta(z',x') \quad \text{for } t > 0,\ z,z' \in Z_t,\ x,x' \in X,$$

if and only if

$$\delta(z_0, x_1 \circ x_1' \ldots x_t \circ x_t') = \delta(z_0,x_1 \ldots x_t) \square \delta(z_0,x_1' \ldots x_t') \tag{4}$$

for $x_1,\ldots,x_t,\ x_1',\ldots,x_t' \in X$.

Let $z = \delta(z_0,p)$, $z' = \delta(z_0,p')$, $p = x_1 \ldots x_t$, $p' = x_1' \ldots x_t'$.

a. (Eq. (4) implies Eq. (7)). $\delta(z,x) \square \delta(z',x') =$

$$= \delta(\delta(z_0,p),x) \square \delta(\delta(z_0,p'), x') = \delta(z_0,px) \square \delta(z_0,p'x') =$$

$$= \delta(z_0,px \circ p'x') = \delta(\delta(z_0,p \circ p'), x \circ x') =$$

$$= \delta(\delta(z_0,p) \square \delta(z_0,p'), x \circ x') = \delta(z \square z', x \circ x').$$

b. (Eq. (7) implies (4)). The proof is by induction on t.
With $z_0 \square z_0 = z_0$ we get from Eq. (7) $\delta(z_0,x_1 \circ x_1') = \delta(z_0,x_1) \square \delta(z_0,x_1')$.
Let for $t > 0$ $\delta(z_0,p \circ p') = \delta(z_0,p) \square \delta(z_0,p')$. Then we conclude

$$\delta(z_0,px_{t+1} \circ p'x_{t+1}') = \delta(\delta(z_0,p \circ p'), x_{t+1} \circ x_{t+1}') =$$

$$= \delta(z \square z',x_{t+1} \circ x_{t+1}') = \delta(z,x_{t+1}) \square \delta(z',x_{t+1}') =$$

$$= \delta(z_0,px_{t+1}) \square \delta(z_0,p'x_{t+1}'). \blacksquare$$

Denoting $x \circ x' = F(x,x')$, $x,x' \in X$; $y \triangledown y' = G(y,y')$, $y,y' \in Y$; $z \square z' = H(z,z')$, $z,z' \in Z_t$, $t = 1,2,\ldots$
Eq. (7) and (8) become

$$\delta(H(z,z'), F(x,x')) = H(\delta(z,x), \delta(z',x')) \tag{9}$$

$$\lambda(H(z,z'), F(x,x')) = G(\lambda(z,x), \lambda(z',x')). \tag{10}$$

Eqs. (9) and (10) are well-known as generalized bisymmetry equations (cf. /1/).

Sometimes we suppose the operation o of an (o,$\triangledown$)-superponable automaton to be associative, commutative or reflexive. Then the operations $\triangledown$ and $\square$ have to be associative, commutative or reflexive, too.

Theorem 5. Let $\mathfrak{A} = (X,Y,Z,z_0,\delta,\lambda)$ be a superponable automaton, $\square$ and Z_t, $t > 0$, as defined in Theorem 1, $Y_t = \{y \mid px \in X^t \wedge y = \lambda(\delta(z_0,p),x)\}$.

1. $x \circ x' = x' \circ x$
2. $x \circ (x' \circ x'') = (x \circ x') \circ x''$
3. $x \circ x = x$,

then for all $t > 0$, $y, y', y'' \in Y_t$, $z, z', z'' \in Z_t$

1. $y \triangledown y' = y' \triangledown y$
   $z \,\square\, z' = z' \,\square\, z$
2. $y \triangledown (y' \triangledown y'') = (y \triangledown y') \triangledown y''$
   $z \,\square\, (z' \,\square\, z'') = (z \,\square\, z') \,\square\, z''$
3. $y \triangledown y = y$
   $z \,\square\, z = z$.

The proof is trivial, but we remark that symmetry, associativity or reflexivity of the operation $\triangledown$ do not imply these properties of the operation o.

The following theorem deals with the existence of identity elements and inverse elements of the operations o, $\triangledown$ and $\square$ of $(o,\triangledown)$-superponable automata.

<u>Theorem 6.</u> Let $\mathfrak{a} = (X,Y,Z,z_o,\delta,\lambda)$ be an $(o,\triangledown)$-superponable automaton and $\square$, Z_t, Y_t as before.

1. If e is the unity element of o (i. e. $x \circ e = e \circ x = x$, $x \in X$), then
   a) $\varepsilon_t = (z_o, \underbrace{e \ldots e}_{t})$ is the unity element of $\square$ restricted to Z_t.
   (i. e. $\varepsilon_t \,\square\, z = z \,\square\, \varepsilon_t = z$, $z \in Z_t$) and
   b) $E_t = g_t(\underbrace{e \ldots e}_{t})$ is the unity element of $\triangledown$ restricted to Y_t.
   (i. e. $E_t \triangledown y = y \triangledown E_t = y$, $y \in Y_t$).
2. If x^{-1} is the inverse element of x, (i. e. $x \circ x^{-1} = e$, $x \in X$) and if $z \in Z_t$, $z = \delta(z_o, x_1 \ldots x_t)$, $y \in Y_t$, $y = g_t^{\mathfrak{a}}(\tilde{x}_1, \ldots, \tilde{x}_t)$, then
   a) $z^{-1} = \delta(z_o, x_1^{-1} \ldots x_t^{-1})$ is the unique inverse element of z,
   $(z \,\square\, z^{-1} = z^{-1} \,\square\, z = \varepsilon_t)$
   b) $y^{-1} = g_t^{\mathfrak{a}}(\tilde{x}_1^{-1}, \ldots, \tilde{x}_t^{-1})$ is the unique inverse element of y,
   $(y \triangledown y^{-1} = y^{-1} \triangledown y = E_t)$.

Proof 1. a) Let $z \in Z_t$, $z = \delta(z_o, x_1 \dots x_t)$. Then $z \,\Box\, \varepsilon_t =$
$\delta(z_o, x_1 \dots x_t) \,\Box\, \delta(z_o, \underbrace{e \dots e}_{t}) = \delta(z_o, x_1 \circ e \dots x_t \circ e) =$
$= \delta(z_o, x_1 \dots x_t) = z = \delta(z_o, e \circ x_1 \dots e \circ x_t) =$
$= \delta(z_o, e \dots e) \,\Box\, \delta(z_o, x_1 \dots x_t) = \varepsilon_t \,\Box\, z$, i. e. ε_t is a unity element of $\Box$ on Z_t.

Let ε_t, $\varepsilon_t^* \in Z_t$ be unity elements. Then we have $\varepsilon_t \,\Box\, \varepsilon_t^* = \varepsilon_t^* = \varepsilon_t$, i. e. ε_t is uniquely determined.

The proof of the other statements of Theorem 6 is quite similar. ■

As an example let us consider the automaton

$$\mathfrak{A} = (X, Y, Z, z_o, \delta, \lambda) \text{ with } X = \{0,1\},\ Y = \{0,1,2,3\}$$

$$Z = \left\{ \begin{pmatrix}0\\0\end{pmatrix}, \begin{pmatrix}0\\1\end{pmatrix}, \begin{pmatrix}1\\0\end{pmatrix}, \begin{pmatrix}1\\1\end{pmatrix} \right\},\ z_o = \begin{pmatrix}0\\0\end{pmatrix}$$

$$z(t+1) = \begin{pmatrix}0 & 1\\1 & 1\end{pmatrix} \cdot z(t) + \begin{pmatrix}1\\0\end{pmatrix} \cdot x(t) \pmod 2 \qquad (11)$$

and

$$y(t) = \begin{cases} 0 & \text{if } u(t) = 0 \text{ and } t \text{ is odd} \\ 1 & \text{if } u(t) = 1 \text{ and } t \text{ is odd} \\ 2 & \text{if } u(t) = 0 \text{ and } t \text{ is even} \\ 3 & \text{if } u(t) = 1 \text{ and } t \text{ is even} \end{cases}$$

with

$$u(t) = (11) \cdot z(t) \quad (\text{mod } 2)) \qquad (12)$$

$\mathfrak{A}$ is a modified linear automaton, and $\mathfrak{A}$ is obviously $(+,\triangledown)$-superponable, where + denotes the usual addition modulo 2 and $\triangledown$ is given by Table 1.

$\triangledown$	0	1	2	3
0	0	1		
1	1	0		
2			2	3
3			3	2

Table 1

The operation $\triangledown$ is only partially determined and the empty squares in Table 1 can be arbitrarily filled up with elements of Y.
Y_t and E_t are determined by

$$Y_{2\mathcal{T}} = \{2,3\},\ E_{2\mathcal{T}} = 2 = g^{\mathfrak{A}}_{2\mathcal{T}}(\underbrace{0,\ldots,0}_{2\mathcal{T}})$$

$$Y_{2\mathcal{T}-1} = \{0,1\},\ E_{2\mathcal{T}-1} = 0 = g^{\mathfrak{A}}_{2\mathcal{T}-1}(\underbrace{0,\ldots,0}_{2\mathcal{T}-1}),\quad \mathcal{T} = 1,2,\ldots$$

On the other hand we have $Z_t = Z$ for $t \geq 2$, $\mathcal{E}_t = \mathcal{E} = \binom{0}{0}$ for $t \geq 1$.

Thus, at different times t, t' we may have different E_t, $E_{t'}$ and in other cases different $\mathcal{E}_t$, $\mathcal{E}_{t'}$, respectively.

At the end of this part we remark that a formulation of superponablility in terms of algebra as homomorphisms and commutative diagrams is possible (cf. /27, /28/) but we do not consider it.

3.2. Binary superponable automata

In the case of binary automata it is possible to completely characterize all (F,G)-superponable automata[1]. An automaton $\mathfrak{A} = (X,Y,Z,z_0,\delta,\lambda)$ is called a binary automaton if $X = Y = \{0,1\}$. $F : X^2 \to X$ and $G : F^2 \to Y$ are 2-ary Boolean functions and $g_t : X^t \to Y$ is a t-ary Boolean function, $t > 0$.

Regarding Definition 1' the automaton $\mathfrak{A}$ is (F,G)-superponable if and only if for all $t > 0$, $x_1,\ldots,x_t$ $x'_1,\ldots,x'_t \in X$

$$g^{\mathfrak{A}}_t(F(x_1,x'_1),\ldots,F(x_t,x'_t)) = G(g^{\mathfrak{A}}_t(x_1,\ldots,x_t),\ g^{\mathfrak{A}}_t(x'_1,\ldots,x'_t)) \quad (2).$$

Firstly we describe the class of binary automata being F-superponable (i. e. (F,F)-superponable) for every 2-ary Boolean function. For this purpose we introduce the notion "selecting automaton".

Definition 2. The automaton $\mathfrak{A} = (X,Y,Z,z_0,\delta,\lambda)$ is called a selecting automaton if for all $t > 0$

$$g^{\mathfrak{A}}_t(x_1,\ldots,x_t) = x_{\mathcal{T}(t)},\qquad \mathcal{T}(t) \in \{1,\ldots,t\}. \tag{13}$$

With this definition we formulate

---

1 The following considerations are based on /17, 15, 12, 13/. In /36/ so-called disjunctive automata superponable with respect to the Boolean function disjunction are introduced.

Theorem 7. Let $\mathfrak{A} = (X,Y,Z,z_0,\delta,\lambda)$ be a binary automaton. Then $\mathfrak{A}$ is F-superponable with respect to every 2-ary Boolean function if and only if $\mathfrak{A}$ is a selecting automaton.
There is no binary automaton $\mathfrak{A}$ superponable with respect to all pairs of Boolean functions F and G.

Proof. a) If $\mathfrak{A}$ is a selecting automaton, then $\mathfrak{A}$ is obviously F-superponable.

b) Similar as in the proof of Theorem 8 $g_t^{\mathfrak{A}}$ is represented in its antivalent normal form, for which the coefficients are to be determined. (For details s. /17/.)

The second statement is a direct consequence of the first. ■

In the following we are mainly interested in "non trivial" functions and automata, which will be called essential functions and essential automata, respectively.

Definition 3. An n-ary Boolean function $\varphi(x_1,\dots,x_n)$, $n \geq 2$, is called non-essential if there exists a unary Boolean function Ψ such that $\varphi(x_1,\dots,x_n) = \Psi(x_j)$, $j \in \{1,\dots,n\}$; and essential otherwise. A binary automaton $\mathfrak{A} = (X,Y,Z,z_0,\delta,\lambda)$ is called non-essential if for every $t > 0$ the Boolean function $g_t^{\mathfrak{A}} : X^t \rightarrow Y$ is non-essential and essential otherwise (cf. ch. 1).

The characterization of all essential binary automata that are (F,G)-superponable with respect to essential Boolean functions F and G is based on the fact that every Boolean function can be uniquely represented in its antivalent normal form, which is sometimes known as Shegalkin's polynomial /39, 5/.

Theorem 8. Let $\varphi(x_1,\dots,x_n)$ be an n-ary Boolean function, $n \geq 0$. Then φ can be uniquely represented as

$$\varphi(x_1,\dots,x_n) = a_0 + \sum_{1\leq i\leq t} a_i \cdot x_i + \sum_{1<i<j\leq t} a_{ij}, x_i \cdot x_j + \dots \qquad (14)$$

$$\dots + a_{12\dots t}\, x_1 \cdot x_2 \cdot \dots \cdot x_t$$

with $a_0, a_1,\dots,a_t,\dots,a_{12\dots t} \in \{0,1\}$. The addition in GF(2) is denoted by +, i. e. addition modulo 2 (antivalence), . denotes the multiplication (or conjunction $\wedge$ /39/).

In the following the symbol . will be omitted.

Thus every 2-ary Boolean function $F(x_1,x_2)$ can be represented as

$$F(x_1,x_2) = a_o + a_1x_1 + a_2x_2 + a_{12}x_1x_2, \qquad (15)$$
$$a_o,\ a_1,\ a_2,\ a_{12} \in \{0,1\}.$$

Examples are $x_1 \vee x_2 = x_1 + x_2 + x_1x_2$ (disjunction), $x_1 \wedge x_2 = x_1x_2$ (conjunction), $x_1 \rightarrow x_2 = 1 + x_1 + x_1x_2$ (implication).
Table 2 shows all $2^{2^2} = 16$ 2-ary Boolean functions and the coefficients a_o, a_1, a_2, a_{12} of their antivalent normal forms (15).

	F_o	F_1	F_2	F_3	F_4	F_5	F_6	F_7	F_8	F_9	F_{10}	F_{11}	F_{12}	F_{13}	F_{14}	F_{15}
a_o	0	1	0	1	0	1	0	1	0	1	0	1	0	1	0	1
a_1	0	0	1	1	0	0	1	1	0	0	1	1	0	0	1	1
a_2	0	0	0	0	1	1	1	1	0	0	0	0	1	1	1	1
a_{12}	0	0	0	0	0	0	0	0	1	1	1	1	1	1	1	1
symbol	0	1	x_1	x_1	x_2	x_2	+	≡	∧			→			∨	
x_1 x_2																
0 0	0	1	0	1	0	1	0	1	0	1	0	1	0	1	0	1
1 0	0	1	1	0	0	1	1	0	0	1	1	0	0	1	1	0
0 1	0	1	0	1	1	0	1	0	0	1	0	1	1	0	1	0
1 1	0	1	1	0	1	0	0	1	1	0	0	1	0	1	1	0

Table 2

Table 2 represents all 2-ary Boolean functions $F(x_1,x_2)$ and the coefficients of their antivalent normal form $F(x_1,x_2) = a_o + a_1x_1 + a_2x_2 + a_{12}x_1x_2$. The functions $F_o,\dots,F_5$ are non-essential, the functions $F_6,\dots,F_{15}$ are essential.

The following theorem deals with the characterization of all essential binary automata superponable with respect to pairs of essential Boolean functions.

<u>Theorem 9.</u> Let $\mathfrak{a} = (X,Y,Z,z_o,\delta,\lambda)$ be an essential binary automaton and let F and G be 2-ary essential Boolean functions. Then $\mathfrak{a}$ is (F,G)-superponable with respect to

1. $F(x_1,x_2) = x_1 + x_2$

$G(x_1,x_2) = x_1 + x_2$

2. $F(x_1,x_2) = x_1 + x_2$

$G(x_1,x_2) = 1 + x_1 + x_2 = x_1 \equiv x_2$

3. $F(x_1,x_2) = 1 + x_1 + x_2 = x_1 \equiv x_2$

$G(x_1,x_2) = x_1 + x_2$

4. $F(x_1,x_2) = 1 + x_1 + x_2 = x_1 \equiv x_2$

$G(x_1,x_2) = 1 + x_1 + x_2 = x_1 \equiv x_2$

5. $F(x_1,x_2) = x_1 + x_2 + x_1x_2 = x_1 \vee x_2$

$G(x_1,x_2) = x_1 + x_2 + x_1x_2 = x_1 \vee x_2$

6. $F(x_1,x_2) = x_1 + x_2 + x_1x_2 = x_1 \vee x_2$

$G(x_1,x_2) = x_1x_2 = x_1 \wedge x_2$

7. $F(x_1,x_2) = x_1x_2 = x_1 \wedge x_2$

$G(x_1,x_2) = x_1 + x_2 + x_1x_2 = x_1 \vee x_2$

8. $F(x_1,x_2) = x_1x_2 = x_1 \wedge x_2$

$G(x_1,x_2) = x_1x_2 = x_1 \wedge x_2$

if and only if we have for all $t > 0$, $x_1,\ldots,x_t \in X$

1. $g_t^{\alpha}(x_1,\ldots,x_t) = \sum_{i=1}^{t} c_i^t x_i$

2. $g_t^{\alpha}(x_1,\ldots,x_t) = 1 + \sum_{i=1}^{t} c_i^t x_i$

3. $g_t^{\alpha}(x_1,\ldots,x_t) = c_0^t + \sum_{i=1}^{t} c^t x_i$ with $c_o^t = \sum_{i=1}^{t} c_i^t$

4. $g_t^{\alpha}(x_1,\ldots,x_t) = c_o^t + \sum_{i=1}^{t} c^t x_i$ with $c_o^t = 1 + \sum_{i=1}^{t} c_i^t$

5. $g_t^{\alpha}(x_1,\ldots,x_t) = c_o^t \vee \bigvee_{i=1}^{t} c_i^t x_i$

6. $g_t^{\alpha}(x_1,\ldots,x_t) = \overline{(c_o^t \vee \bigvee_{i=1}^{t} c_i^t x_i)}$

7. $g_t^{\alpha}(x_1,\ldots,x_t) = \overline{(c_o^t \wedge \bigwedge_{i\in I(t)} x_i)} = 1 + c_o^t \prod_{i=1}^{t} (1 + c_i^t + c_i^t x_i)$

with $I(t) \subseteq \{1,\ldots,t\}$ and $a_i^t = 1$ iff $i \in I(t)$

8. $g_t^{\alpha}(x_1,\ldots,x_t) = c_o^t \wedge \bigwedge_{i\in I(t)} x_i = c_o^t \cdot \prod_{i=1}^{t} (1 + c_i^t + c_i^t x_i)$

with $I(t) \subseteq \{1,\ldots,t\}$ and $c_i^t = 1$ iff $i \in I(t)$.

If F and G are essential functions different from the pairs 1. - 8. and if α is (F,G)-superponable, then α is non-essential.

Proof. α is (F,G)-superponable if and only if for all $t > 0$, $x_1,\ldots,x_t,\ x_1',\ldots,x_t' \in X$

$$g_t^{\alpha}(F(x_1,x_1'),\ldots,F(x_t,x_t')) = G(g_t^{\alpha}(x_1,\ldots,x_t),\ g_t^{\alpha}(x_1',\ldots,x_t')). \quad (2)$$

For $t \geq 2$ we assume $t - 2$ variables and without loss of generality the variables $x_3,\ldots,x_t,\ x_3',\ldots,x_t'$, to be fixed, $x_3 = x_3' = x_3^o,\ldots,x_t = x_t' = x_t^o$; $x_3^o,\ldots,x_t^o \in \{0,1\}$. Denoting $g(x_1,x_2) = g_t(x_1,x_2,x_3^o,\ldots,x_t^o)$, $h(x_1,x_2) = g_t(x_1,x_2,F(x_3^o,x_3^o),\ldots,F(x_t^o,x_t^o))$, Eq. (2) reduces to

$$G(g(x_1,x_2),\ g(x_1',x_2')) = h(F(x_1,x_1'),\ F(x_2,x_2')). \quad (16)$$

F, G, g, h may be uniquely represented as

$$\begin{aligned} F(x_1,x_2) &= A_o + A_1x_1 + A_2x_2 + A_{12}x_yx_2 \\ G(x_1,x_2) &= a_o + a_1x_1 + a_2x_2 + a_{12}x_1x_2 \\ g(x_1,x_2) &= B_o + B_1x_1 + B_2x_2 + B_{12}x_1x_2 \\ h(x_1,x_2) &= b_o + b_1x_1 + b_2x_2 + b_{12}x_1x_2\ . \end{aligned} \quad (17)$$

It is easy to see that the Boolean function $F(x_1,x_2)$ is essential if 1. $A_{12} = 1$ or if 2. $A_{12} = 0$ and $A_1 = A_2 = 1$.

Let F and G be given Boolean functions. Taking into account Eq. (17) we obtain from Eq. (16) a system of 16 nonlinear equations for the coefficients B_o, B_1, B_2, B_{12}, b_o, b_1, b_2, b_{12} of the functions g and h, respectively.

$$\begin{aligned}
&1.\ a_o + a_1 B_o + a_2 B_o + a_{12} = b_o + b_1 A_o + b_2 A_o + b_{12} A_o \\
&2.\ a_1 B_1 + a_{12} B_o B_1 = b_1 A_1 + b_{12} A_o A_1 \\
&3.\ a_1 B_2 + a_{12} B_o B_2 = b_2 A_1 + b_{12} A_o A_1 \\
&4.\ a_2 B_1 + a_{12} B_o B_1 = b_1 A_2 + b_{12} A_o A_2 \\
&5.\ a_2 B_2 + a_{12} B_o B_2 = b_2 A_2 + b_{12} A_o A_2 \\
&6.\ a_1 B_{12} + a_{12} B_o B_{12} = b_{12} A_1 \\
&7.\ a_{12} B_1 = b_1 A_{12} + b_{12} A_o A_{12} \\
&8.\ a_{12} B_1 B_2 = b_{12} A_1 A_2 \\
&9.\ a_{12} B_1 B_2 = b_{12} A_1 A_2 \\
&10.\ a_{12} B_2 = b_2 A_{12} + b_{12} A_o A_{12} \\
&11.\ a_2 B_{12} + a_{12} B_o B_{12} = b_{12} A_2 \\
&12.\ a_{12} B_{12} B_1 = b_{12} A_1 A_{12} \\
&13.\ a_{12} B_{12} B_2 = b_{12} A_1 A_{12} \\
&14.\ a_{12} B_{12} B_1 = b_{12} A_2 A_{12} \\
&15.\ a_{12} B_{12} B_2 = b_{12} A_2 A_{12} \\
&16.\ a_{12} B_{12} = b_{12} A_{12}
\end{aligned} \tag{18}$$

The equations 8. and 9. corresponding to x_1x_2' and $x_1'x_2$ are identical.

If F and G are the pairs of functions listed from 1. to 8., we have

1. $g(x_1,x_2) = B_o + B_1x_1 + B_2x_2$
   $h(x_1,x_2) = B_1x_1 + B_2x_2$

2. $g(x_1,x_2) = B_o + B_1x_1 + B_2x_2$
   $h(x_1,x_2) = 1 + B_1x_1 + B_2x_2$

3. $g(x_1,x_2) = B_o + B_1x_1 + B_2x_2$

$$h(x_1,x_2) = B_1 + B_2 + B_1x_1 + B_2x_2$$

4. $g(x_1,x_2) = B_o + B_1x_1 + B_2x_2$

$$h(x_1,x_2) = 1 + (B_1 + B_2) + B_1x_1 + B_2x_2$$

5.

$$g(x_1,x_2) = h(x_1,x_2) = \begin{cases} B_o \\ x_1 \\ x_2 \\ x_1 + x_2 + x_1x_2 \end{cases}$$

6.

$$g(x_1,x_2) = h(x_1,x_2) = \begin{cases} B_o \\ 1 + x_1 \\ 1 + x_2 \\ 1 + x_1 + x_2 + x_1x_2 = x_1 \quad x_2 \end{cases}$$

7.

$$g(x_1,x_2) = h(x_1,x_2) = \begin{cases} B_o \\ 1 + x_1 \\ 1 + x_2 \\ 1 + x_1x_2 \end{cases}$$

8.

$$g(x_1,x_2) = h(x_1,x_2) = \begin{cases} B_o \\ x_1 \\ x_2 \\ x_1x_2 \end{cases}$$

with $B_o,B_1,B_2 \in \{0,1\}$.

If (F,G) is a pair of essential Boolean functions not listed from 1. - 8. and if $g(x_1,x_2)$ is a solution of Eq. (16), then $g(x_1,x_2)$ is non-essential and hence $g_t(x_1,\dots,x_t)$ is non-essential, too.

For given $g(x_1,x_2)$ and $h(x_1,x_2)$ 1. to 8. we have to determine now the corresponding functions $g_t(x_1,\dots,x_t)$, $t > 0$. Explicitly this will be done in the cases 2. $F(x_1,x_2) = x_1 + x_2$, $G(x_1,x_2) = 1 + x_1 + x_2 = x_1 \equiv x_2$ and 5. $F(x_1,x_2) = G(x_1,x_2) = x_1 + x_2 + x_1x_2 = x_1 \vee x_2$. The other cases are similar.

The function $g_t^{\alpha}(x_1,\dots,x_t)$ is supposed to be in its antivalent normal form

$$g_t(x_1,\dots,x_t) = c_o^t + \sum_{1\le i\le t} c_i^t\, x_i + \sum_{1\le i<j\le t} c_{ij}^t\, x_ix_j + \dots \\ \dots + c_{12\dots t}^t\, x_1x_2 \dots x_t\,. \tag{19}$$

Let $F(x_1,x_2) = x_1 + x_2$ and $G(x_1,x_2) = 1 + x_1 + x_2$ (case 2). Then we have

$$g(x_1,x_2) = B_o + B_1x_1 + B_2x_2 = g_t^{\alpha}(x_1,x_2,x_3^o,\dots,x_t^o) \tag{20}$$

$$h(x_1,x_2) = 1 + B_1x_1 + B_2x_2 = g_t^{\alpha}(x_1,x_2,F(x_3^o,x_3^o),\dots,F(x_3^o,x_3^o)) \tag{21}$$

$$B_o,B_1,B_2 \in \{0,1\} .$$

Since $F(x,x) = 0$, $x \in X$, Eq. (21) becomes

$$1 + B_1x_1 + B_2x_2 = g_t(x_1,x_2,0,\dots,0) = c_o + c_1x_1 + c_2x_2 + c_{12}x_1x_2$$

and hence $c_o = 1$.

Let $c_{i_1\dots i_k} = 1$ and k minimal. Without loss of generality let $c_{1\dots k} = 1$.

Then $g_t^{\alpha}(x_1,x_2,\underbrace{1\ \dots\ 1}_{k-2},0\ \dots\ 0) = d_o + d_1x_1 + d_2x_2 + x_1x_1,\quad d_o,d_1,d_2 \in \{0,1\}$

in contradiction to $g(x_1,x_2) = B_o + B_1x_1 + B_2x_2$.

Consequently we have

$$g_t^{\alpha}(x_1,\dots,x_t) = 1 + \sum_{1\le i\le t} c_i^t\, x_i . \tag{22}$$

Finally it is easy to check that an automaton $\mathfrak{A}$ with g_t^{α} defined by Eq. (22) is (F,G)-superponable with $F(x_1,x_o) = x_1 + x_2$ and $G(x_1,x_2) = 1 + x_1 + x_2$.

Now let $F(x_1,x_2) = G(x_1,x_2) = x_1 + x_2 + x_1x_2 = x_1 \vee x_2$ (case 5).

Since $F(x,x) = x$, $x \in X$, we have $g_t(x_1,x_2,x_3^o,\dots,x_t^o) = g_t(x_1,x_2,F(x_3^o,x_3^o),\dots,F(x_t^o,x_t^o)) = g(x_1,x_2) = h(x_1,x_2)$ with

$$\begin{aligned} &g_1(x_1,x_2) = h_1(x_1,x_2) = 0, \quad g_2(x_1,x_2) = h_2(x_1,x_2) = 1\\ &g_3(x_1,x_2) = h_3(x_1,x_2) = x_1 \quad g_4(x_1,x_2) = h_4(x_1,x_2) = x_2\\ &g_5(x_1,x_2) = h_5(x_1,x_2) = x_1 + x_2 + x_1x_2 = x_1 \vee x_2 . \end{aligned} \tag{23}$$

The function g_t^{α} is again supposed to be in its antivalent normal form defined by Eq. (19).

<u>a.</u> Let $c_o = 1$. We suppose $c_{i_1\dots i_k} = 1$ with $k \ge 1$ and k minimal. Without loss of generality let $c_{1\dots k} = 1$. If $k = 1$, we obtain

$g_t(x_1,x_2,\underbrace{1,\dots,1}_{k-2},0,\dots,0) = 1 + x_1x_2$ in contradiction to Eq. (23).

Thus, if $c_o = 1$, all the other coefficients of the antivalent normal form have to be zero.

b. Let $c_o = 0$. Without loss of generality let $a_1 = \dots = a_k = 1$, $a_{k+1} = \dots = a_t = 0$, $0 < k < t$. Then we show

$$a_{i_1\dots i_l} = 0 \quad \text{if} \quad i_l > k, \ (l \geq 2) \tag{24}$$

We suppose that Eq. (24) is not true. Let $a_{m+1\dots m+l} = 1$, $l \geq 2$, $m + l > k$ and l minimal. Then we have $g_t^{\alpha}(0,\dots,0,x_{m+1},1,\dots,1,x_{m+l},0,\dots,0) = d_1x_{m+1} + x_{m+1}\, x_{m+l} + d_2$, $d_1,d_2 \in \{0,1\}$ in contradiction to Eq. (23).

Thus g_t^{α} does not depend on $x_{k+1},\dots,x_t$.

Now we show:

$$a_{i_1\dots i_l} = 1 \quad \text{if} \quad i_l \leq k, \ (l \geq 2). \tag{25}$$

We suppose that Eq. (25) is not true. Let $k > 2$ and let without loss of generality $a_{1\dots l} = 0$, $2 \leq l \leq k$, and l minimal.

If $l = 2$, then $g_t^{\alpha}(x_1,x_2,0,\dots,0) = x_1 + x_2$ in contradiction to Eq. (23).

If $l > 2$, we have $g_t(x_1,x_2,\underbrace{1,\dots,1}_{l-2},0,\dots,0) =$

$$= \sum_{i=1}^{l-2} \binom{l-2}{i} + (x_1 + x_2) \sum_{i=0}^{l-2} \binom{l-2}{i} + x_1x_2 \sum_{i=0}^{l-3} \binom{l-2}{i} = 1 + x_1x_2 \pmod 2$$

in contradiction to Eq. (23).

Thus we have with $I = \{i_1,\dots,i_k\} \subseteq \{1,\dots,t\}$.

$$g_t^{\alpha}(x_1,\dots,x_t) = \left\{ \sum_{i\in I}^{1} x_i + \sum_{\substack{i<j\\ i,j\in I}} x_ix_j + \sum_{\substack{i<j<k\\ i,j,k\in I}} x_ix_jx_k + \dots + x_{i_1}x_{i_2} \dots \right.$$
$$\dots x_{i_k} = \bigvee_{i\in I} x_i$$

Till now we have shown:

Let α be $(\vee,\vee)$-superponable, then g_t^{α} has to be of the form

$$g_t(x_1,\dots,x_t) = c_o^t \vee \bigvee_{i=1}^{t} c_i^t\, x_i, \quad c_i \in \{0,1\}. \tag{26}$$

On the other hand it is easy to check that a binary automaton with g_+^{α}

defined by Eq. (27) is $(\vee,\vee)$-superponable.

Summarizing we have proved:

Let $\mathfrak{A}$ be a binary automaton. Then $\mathfrak{A}$ is $(\vee,\vee)$-superponable if and only if $g_t^{\mathfrak{A}}$ can be expressed as

$$g_t(x_1,\ldots,x_t) = c_o^t \vee \bigvee_{i=1}^{t} c_i^t\, x_i .$$

In a similar way the other statements of Theorem 9 can be proved. ■

Now we extend the description of the input-output-behaviour by means of impulse responses to (F,G)-superponable binary automata.

<u>Theorem 10.</u> Let $\mathfrak{A} = (X,Y,Z,z_o,\delta,\lambda)$ be a binary (F,G)-superponable automaton, and let for every $x \in X$: $F(e,x) = F(x,e) = x$ with $e \in \{0,1\}$. Then for every input sequence $x_1\ldots x_n \in X^n$, $n \geq 1$, the corresponding output sequence $\lambda(x_1\ldots x_n)$ is uniquely determined by the n + 1 output sequences $\lambda(e\ldots e)$, $\lambda(\bar{e}\, e\ldots e),\ldots,$ $\lambda(e\ldots e\, \bar{e})$ and can be computed by

$$\lambda(x_1\ldots x_n) = \lambda(x_1 e\ldots e)G\, \lambda(ex_2\ldots e)G\, \ldots\, G\, \lambda(e\ldots ex_n). \qquad (27)$$

Proof. Let for $x \in X$: $F(x,e) = F(x,e) = x$. Then every $x_1\ldots x_n \in X^n$, $n \geq 1$, can be expressed as

$$x_1\ldots x_n = (x_1 e\ldots e)F(ex_2 e\ldots e)F\, \ldots\, F(e\ldots ex_n). \qquad (28)$$

Since $\mathfrak{A}$ is supposed to be (F,G)-superponable, we obtain from Eq. (28) Eq. (27). Because of $x_i \in \{e,\bar{e}\} = \{0,1\}$, $\lambda(e\ldots e)$, $\lambda(\bar{e}\, e\ldots e),\ldots$ $\ldots$, $\lambda(e\ldots e\, \bar{e})$ uniquely determine $\lambda(x_1\ldots x_n)$. ■

In the case of antivalence, equivalence, disjunction, and conjunction we have for $x \in \{0,1\}$

1. $x + 0 = 0 + x = x$
2. $x \equiv 1 = 1 \equiv x = x$
3. $x \vee 0 = 0 \vee x = x$
4. $x \wedge 1 = 1 \wedge x = x$.

Thus we can apply Theorem 10 to all the essential binary automata characterized by Theorem 9 as (F,G)-superponable automata with respect to essential Boolean functions F and G. The following theorem shows that the input-output-behaviour of finite (F,G)-superponable automata is determined by a finite set of periodic generalized impulse responses.

<u>Theorem 11.</u> Let $\mathfrak{A} = (X,Y,Z,z_o,\delta,\lambda)$ be a finite (F,G)-superponable binary automaton with n states, say $Z = \{z_o,z_1,\ldots,z_{n-1}\}$, $n > 0$.

Then we have:

1. For every $k > 0$ there exists a positive integer i_k, $0 \leq i_k < n$ such that

$$\lambda(z_0,\underbrace{e\ldots e}_{k}\ \bar{e}\ e\ e\ldots) = \lambda(z_0,\underbrace{e\ldots e}_{k})\ \lambda(z_{i_k},\bar{e}\ e\ e\ \ldots). \tag{28}$$

2. For every $z \in Z$ exist positive integers N_z and l_z with $0 < N_z \leq n$ and $0 \leq l_z < n$ such that

$$\lambda(z,\bar{e}\ e\ e\ \ldots) = y_1^z\ldots y_z^t\ \underbrace{y^z_{l_z+1}\ldots y^z_{l_z+N_z}}_{\text{periodic}}. \tag{29}$$

3. There exist positive integers N and l with $0 < N \leq n$ and $0 \leq l < r$ such that

$$\lambda(z_0,e\ e\ e\ \ldots) = y_1\ldots y_l\ \underbrace{y_{l+1}\ldots y_{l+N}}_{\text{periodic}}. \tag{30}$$

Proof. For the proof it is only necessary that $\mathfrak{A}$ is a finite automaton with n states. ■

Of special interest is the question under which conditions the input-output-behaviour of an (F,G)-superponable automaton can be determined by a simple generalized impulse response. The answer is given in the next theorem.

Theorem 12. Let $\mathfrak{A} = (X,Y,Z,z_0,\delta,\lambda)$ be an (F,G)-superponable automaton and let $F(x,e) = F(e,x) = x$ for $x \in X$. Then we have for every $k > 0$

$$\lambda(\underbrace{e\ldots e}_{k}\ \bar{e}\ e\ e\ldots) = \underbrace{b\ldots b}_{k}\ \lambda(\bar{e}\ e\ e\ldots) \tag{31}$$

if and only if $\delta(z_0,e) \sim z_0$.

Proof.

a. $\delta(z_0,e) \sim z_0$ implies $\delta(z_0,e\ldots e) \sim z_0$. Hence it follows that $\lambda(\underbrace{e\ldots e}_{k}\ \bar{e}\ e\ e) = \underbrace{b\ldots b}_{k}\ \lambda(\bar{e}\ e\ e\ldots)$ with $b = \lambda(z_0,e)$.

b. Let for every $k > 0$ $\lambda(\underbrace{e\ldots e}_{k}\ \bar{e}\ e\ e\ldots) = \underbrace{b\ldots b}_{k}\ \lambda(\bar{e}\ e\ e\ldots)$

and let $\delta(z_0,e) \nsim z_0$. Then there has to exist an input sequence

$x_1 \ldots x_n$, $n > 0$, such that $\lambda(x_1 \ldots x_n) = y_1 \ldots y_n$ and $\lambda(e\ x_1 \ldots x_n) = b\ y'_1 \ldots y'_n$ with $(y_1 \ldots y_n) \neq (y'_1 \ldots y'_n)$. Since $\mathfrak{a}$ is (F,G)-superponable, we have

$$\lambda(x_1 \ldots x_n) = [\lambda(x_1 e \ldots e)]\ G\ [\lambda(e\ x_2 e \ldots e)]\ G \ldots G\ [\lambda(e \ldots e\ x_n)] = \\ = [\lambda(x_1 e \ldots e)]\ G\ [b\ \lambda(x_2 e \ldots e)]\ G \ldots G\ [b \ldots b\ \lambda(x_n)]$$

and

$$\lambda(ex_1 \ldots x_n) = [\lambda(ex_1 e \ldots e)]\ G\ [\lambda(eex_2 e \ldots e)]\ G \ldots G\ [\lambda(e \ldots ex_n)] \\ = [b\ \lambda(x_1 e \ldots e)]\ G\ [b\ b\ \lambda(x_2 e \ldots e)]\ G \ldots G\ [b \ldots b\ \lambda(x_n)] \\ = c\ \lambda(x_1 \ldots x_n),\ c \in \{0,1\}, \text{ in contradiction to } \\ (y_1 \ldots y_n) \neq (y'_1 \ldots y'_n). \ \blacksquare$$

Thus the input-output-behaviour of an (F,G)-superponable automaton for which the conditions of Theorem 12 are satisfied is completely determined by $b = \lambda(z_o, e)$ and $\lambda(\bar{e}\ e\ e \ldots)$.

Now we discuss the problem of the explicitly state dependent description of (F,G)-superponable automata and their modelling by means of known impulse responses.

As we know, a (binary) linear automaton, here on GF(2), can be defined by its next-state function and its output function in matrix form

$$z(t+1) = A(t)\ z(t) + B(t)\ x(t) \tag{32}$$

$$y(t) = C(t)\ z(t) + D(t)\ x(t) \tag{33}$$

x,y and z are l-, m-, n-dimensional vectors on GF(2). Since the elements of the matrices $A = (a_{ir})$, $B = (b_{ir})$, $C = (c_{js})$, $D = (d_{js})$ are 0 and 1, we can write instead of Eqs. (32) and (33)

$$z_i(t+1) = \sum_{r \in I_i^1(t)} z_r(t) + \sum_{r \in I_i^2(t)} x_r(t) \tag{34}$$

$$y_j(t) = \sum_{s \in I_j^3(t)} z_s(t) + \sum_{s \in I_j^4(t)} x_s(t) \tag{35}$$

with

$$a_{ir} = 1 \text{ iff } r \in I_i^1 \qquad c_{js} = 1 \text{ iff } s \in I_j^3$$

$$b_{j,r} = 1 \text{ iff } r \in I_i^2 \qquad d_{js} = 1 \text{ iff } s \in I_j^4,$$

where

$I_i^1 \subseteq \{1,\dots,n\}$, $I_i^2 \subseteq \{1,\dots,l\}$, $I_j^3 \subseteq \{1,\dots,n\}$, $I_j^4 \subseteq \{1,\dots,l\}$, (36)
$i = 1,\dots,n$; $j = 1,\dots,m$ are index sets.

As a natural generalization of linear automata we introduce so-called disjuntive and conjunctive automata.
A disjunctive automaton $\mathfrak{D}$ is defined by its next-state function

$$z_i(t+1) = \bigvee_{r \in I_i^1(t)} z_r(t) \vee \bigvee_{r \in I_i^2(t)} x_r(t) = \bigvee_r a_{ir}^t\, z_r(t) \vee \bigvee_r b_{ir}^t\, x_r(t) \quad (37)$$

and its output function

$$y_j(t) = \bigvee_{s \in I_j^3(t)} z_s(t) \vee \bigvee_{s \in I_j^4(t)} x_s(t) = \bigvee_s c_{js}^t\, z_s(t) \vee \bigvee_s d_{js}\, x_s(t) \quad (38)$$

(Disjunctive automata are introduced in /36/).

Similarly we define a conjunctive automaton K by its next-state function

$$\begin{aligned} z_i(t+1) &= \bigwedge_{r \in I_i^1(t)} z_r(t) \wedge \bigwedge_{r \in I_i^2(t)} x_r(t) = \\ &= \prod_r (1 + a_{ir}^t + a_{ir}^t\, z_r(t)) \prod_r (1 + b_{ir}^t + b_{ir}^t\, x_r(t)) \end{aligned} \quad (39)$$

and its output function

$$\begin{aligned} y_j(t) &= c_{jo}^t \bigwedge_{s \in I_j^3(t)} z_s(t) \wedge \bigwedge_{s \in I_j^4(t)} x_s(t) = \\ &= c_{jo}^t \prod_s (1 + c_{js}^t + c_{js}^t\, z_s(t)) \prod_s (1 + d_{js}^t + d_{js}^t\, x_s(t)). \end{aligned} \quad (40)$$

As in the case of linear automata we have:

$a_{ir}^t = 1$ iff $r \in I_i^1(t)$, $b_{ir}^t = 1$ iff $r \in I_i^2(t)$,

$c_{js}^t = 1$ iff $s \in I_i^3(t)$, $d_{js} = 1$ iff $s \in I_j^4(t)$ and $c_{oj}^t \in \{0,1\}$.

The addition in (39) is modulo 2.

The index sets are supposed to be restricted by condition (36), where l and m are assumed to be finite, n may be infinite. If the index sets are empty, the corresponding disjunctions are supposed to be zero and the

corresponding conjunctions 1.
In the following we also need the concept of negated linear, negated disjunctive and negated conjunctive automata.

In general we define

Definition 3. Let $\mathcal{A} = (X,Y,Z,z_0,\delta,\lambda)$ be a binary automaton. Then $\bar{\mathcal{A}} = (X,Y,Z,z_0,\delta,\bar{\lambda})$ with $\bar{\lambda}(z,x) = \overline{\lambda(z,x)} = 1 + \lambda(z,x)$ is called the negated automaton of $\mathcal{A}$. (The addition is modulo 2.)

Thus a negated linear automaton $\bar{\mathcal{L}}$ on GF(2) is defined by Eq. (34) (next-state function) and the output function

$$y_j(t) = 1 + \sum_{s\in I_j^3(t)} z_s(t) + \sum_{s\in I_j^4(t)} x_s(t) \tag{41}$$

A negated linear automaton is linear.
A negated disjunctive automaton $\bar{\mathcal{D}}$ is defined by Eq. (37) (next-state function) and the output function

$$y_j(t) = 1 + \bigvee_{s\in I_j^3(t)} z_s(t) \vee \bigvee_{s\in I_j^4(t)} x_s(t) \tag{42}$$

and a negated conjunctive automaton $\bar{\mathcal{K}}$ is defined by Eq. (39) (next-state function) and the output function

$$y_j = 1 + c_{oj}^t \bigwedge_{s \in I_j^3(t)} z_s(t) \vee \bigwedge_{s \in I_j^4(t)} x_s(t) \tag{43}$$

For a simple presentation we restrict ourselves in the following to automata with 1-dimensional input and output only, i. e. $l = m = 1$. The next theorem deals with the state-dependent description of (F,G)-superponable automata.

Theorem 13. Let $\mathcal{A} = (X,Y,Z,z_0,\delta,\lambda)$ be a 1. $(+,+)$, 2. $(+,\equiv)$-, 3. $(\equiv,+)$-, 4. $(\equiv,\equiv)$-, 5. $(\vee,\vee)$-, 6. $(\vee,\wedge)$-, 7. $(\wedge,\vee)$-, 8. $(\wedge,\wedge)$-superponable binary automaton. Then $\mathcal{A}$ is equivalent to a

1. linear automaton with $\lambda(0\ 0\ 0\ \ldots) = 0\ 0\ 0\ \ldots$
2. linear automaton with $\lambda(0\ 0\ 0\ \ldots) = 1\ 1\ 1\ \ldots$
3. linear automaton with $\lambda(1\ 1\ 1\ \ldots) = 0\ 0\ 0\ \ldots$
4. linear automaton with $\lambda(1\ 1\ 1\ \ldots) = 1\ 1\ 1\ \ldots$
5. disjunctive automaton

6. negated disjunctive automaton
7. negated conjunctive automaton
8. conjunctive automaton.

If $\mathfrak{A}$ is finite, the automata listed from 1. - 8. can be constructed as a finite periodically time-variant automaton.

Proof. The proof is performed for $(+,\equiv)$- and $(\vee,\wedge)$-superponable automata. The other cases are quite similar.

1. Let $\mathfrak{A}$ be $(+,\equiv)$-superponable. Then, by Theorem 9 we have

$y_t = g_t^{\mathfrak{A}}(x_1,\dots,x_t) = 1 + \sum_{i=1}^{t} c_i^t\, x_i$. It is easy to see that the automaton $\mathcal{L}_1$ represented in Fig. 3 has the desired input-output-behaviour. The initial state of the delay element 1 is 1, the initial state of all the other delay elements is zero. The dimension of the state vectors is infinite.

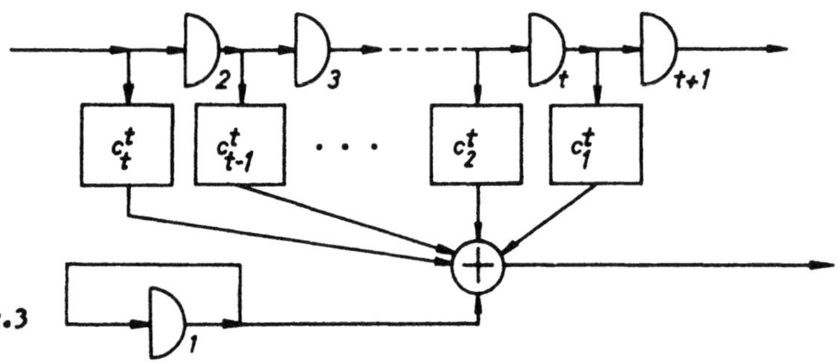

Fig.3

The explicit state-dependent form of $\mathcal{L}_1$ is

$$z_1(t+1) = z_1(t) = 1$$
$$z_2(t+1) = x(t)$$
$$z_{n+1}(t+1) = z_n(t) \qquad n > 2 \tag{44}$$

$$y(t) = \sum_{j=1}^{t-1} c_{t-j}^t\, z_{j+1} + c_t^t\, x_t \tag{45}$$

$\mathcal{L}_1$ is a linear automaton. Obviously we have $\lambda(0\ 0\ 0\ \dots) = 1\ 1\ 1\ \dots$

2. Let $\mathfrak{A}$ be $(\vee,\wedge)$-superponable. By Theorem 9 we have

$y_t = g_t^{\mathfrak{a}}(x_1,\dots,x_t) = \overline{c_0^t \vee \bigvee_{i=1}^{t} c_i^t x_i}$, and it is easy to see that the automaton $\mathfrak{D}_1$ represented in Fig. 4 has the desired input-output-behaviour.

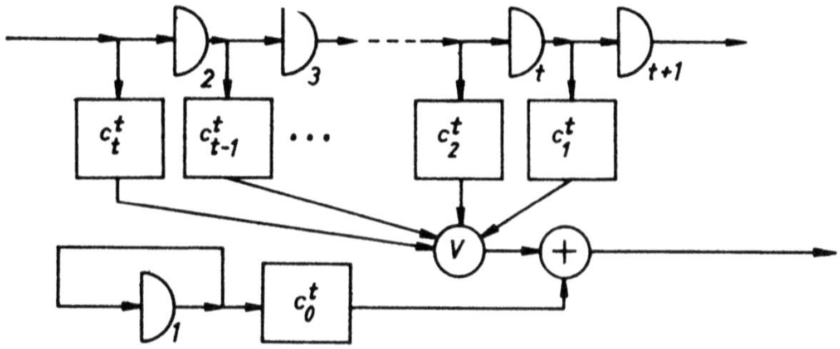

Fig.4

The delay element 1 is in initial state 1, all the other delay elements are in the initial state 0.

The explicitly state dependent form of $\mathfrak{D}_1$ is

$$z_1(t+1) = z_1(t+1) = 1,$$
$$z_2(t+1) = x(t),$$
$$z_{n+1}(t+1) = z_n(t), \qquad n > 2 \tag{46}$$

$$y(t) = 1 + c_0^t\, z_1(t) \vee \bigvee_{j=1}^{t-1} c_{t-j}^t\, z_{j+1}. \tag{47}$$

We show now: Let $\mathfrak{a}$ be a finite $(\vee,\wedge)$-superponable automaton, then $\mathfrak{a}$ is equivalent to a finite periodically time-variant negated disjunctive automaton.

By Theorem 10 we know that the input-output-behaviour of $\mathfrak{a}$ is determined by its generalized impulse responses $\lambda(0\ 0\ 0\ \dots)$, $\lambda(1\ 0\ 0\ \dots)$, $\lambda(0\ 1\ 0\ 0\ \dots)$, $\lambda(0\ 0\ 1\ 0\ \dots), \dots$

First we realize the $(\vee,\vee)$-superponable negated automaton $\bar{\mathfrak{a}}$ with the generalized inpulse responses $\bar\lambda(0\ 0\ 0\ \dots)$, $\bar\lambda(1\ 0\ 0\ \dots)$, $\bar\lambda(0\ 1\ 0\ \dots), \dots$

Let $Z = \{z_0,\dots,z_{n-1}\}$, $n > 0$ and let $z_t^o = \delta(z_o, \underbrace{0\ \dots\ 0}_{t})$.

Then there have to exist numbers i and m; $0 \le i,\ m < n$; such that

$$z_o z_1^o z_2^o \ldots = z_o \; z_1^o \ldots z_{i-1}^o \underbrace{z_i^o \ldots z_{i+m-1}^o}_{\text{periodic}} \tag{48}$$

and

$$\bar{\lambda}(0\;0\;0\;\ldots) = b_o \; b_1 \ldots b_{i-1} \underbrace{b_i \ldots b_{i+m-1}}_{\text{periodic}} . \tag{49}$$

Every $M > i$ can be uniquely represented as

$$M = i + N\,m + K \quad \text{with} \quad 0 \leq K < m,$$

where N is a non-negative integer, i. e. $K = (M - i) \bmod. m$. Thus we have for every $P > 0$

$$\bar{\lambda}(0\ldots0\;100\ldots) = \begin{cases} \underbrace{0\ldots0}_{p} \; \bar{\lambda}(z_p^o,100\ldots) \vee \lambda(000\ldots) & P \leq i \\ \underbrace{0\ldots0}_{p} \; \bar{\lambda}(z_{i+K}^o(100\ldots) \vee \lambda(000\ldots) & P > i . \end{cases} \tag{50}$$

Since $\bar{\mathfrak{a}}$ is a finite automaton with n states, we conclude that for every $z_k^o \in Z$ there have to exist Q_k, R_k, $n \geq Q_k, R_k > 0$ such that

$$\bar{\lambda}(z_k^o,1\;0\;0\;\ldots) = r_1^k \ldots r_{Q_k}^k \underbrace{s_1^k \ldots s_{R_k}^k}_{\text{periodical}} \tag{51}$$

A $(\vee,\vee)$-superponable automaton $\mathfrak{D}_k$ for which

$$\bar{\lambda}(\underbrace{0\ldots0}_{p}100\ldots) = \underbrace{0\ldots0}_{p} \; \bar{\lambda}(z_k^o,100\ldots) \vee \bar{\lambda}(000\ldots), \tag{52}$$

taking into account Eqs. (49) and (51), for every $P > 0$ is fulfilled, is represented in Fig. 5.

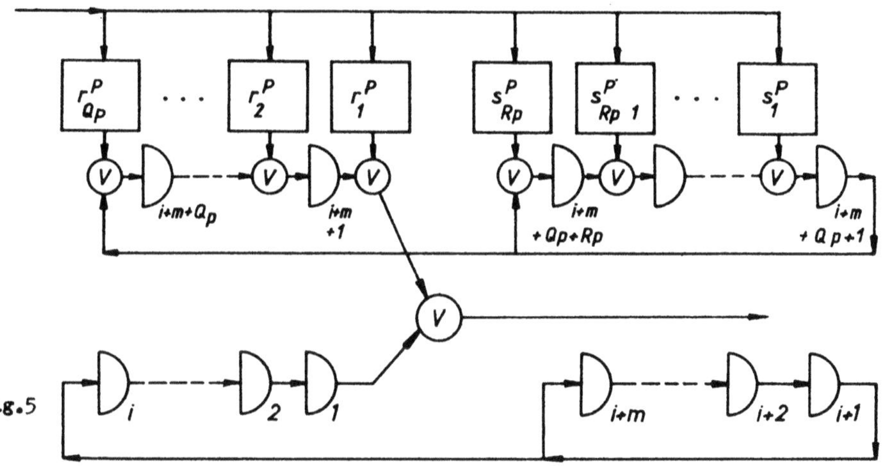

Fig.5

At time t = 0 the delay elements 1,2,...,m+i are in the states $b_0,\dots,b_{i+m-1}$, all the other delay elements are in state 0.

Furtheron we need an autonomous Medwedew-type automaton $\mathcal{C}$ with the state set $Z_{\mathcal{C}} = \{z_0, z_1^0,\dots,z_{i+m-1}^0\}$ and the state diagram of Fig. 6.

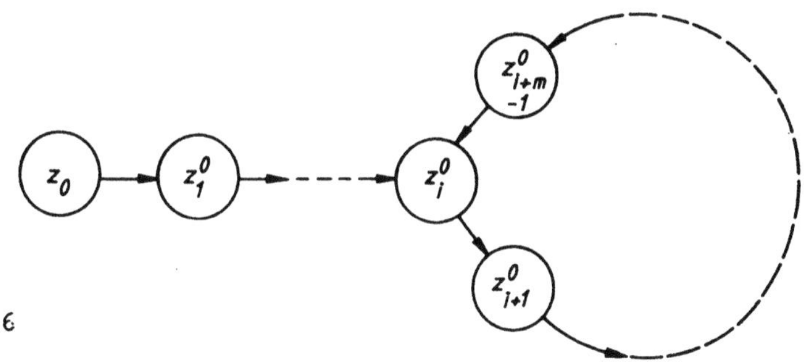

Fig.6

By use of $\mathcal{C}$ and the automata $\mathcal{D}_k$, k = 0,...,i+m-1 we construct the automaton $\mathcal{D}$ as represented in Fig. 7.

The switch S_k is in position 1 iff $\mathcal{C}$ is in the state z_k^0. $\mathcal{D}$ can be interpreted as a finite periodically time-variant disjunctive automaton. For every P > 0 the generalized impulse responses $\lambda(\underbrace{0\dots0}_{P}\ 1\ 0\ 0\ \dots)$ of the negated automaton $\overline{\mathcal{D}}$ are as desired and $\overline{\mathcal{D}}$ is by Theorem 10 equivalent to $\mathcal{A}$. $\overline{\mathcal{D}}$ can be interpreted as a finite periodically time-variant negated disjunctive automaton.

In the other cases the proof is similar. ■

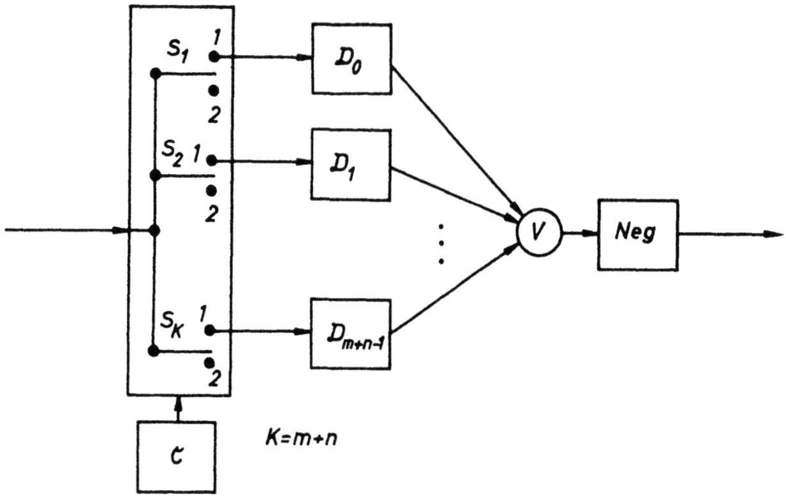

Fig.7

To a certain extend the theory of binary (F,G)-superponable automata seems to be complete. Most of the proofs of the previous theorems are essentially based on the fact that every Boolean function can be uniquely represented in its antivalent normal form, i. e. as a polynomial over GF(2).

Now the question arises how to treat the superponability of automata with finite input- and output sets or even with infinite input- and output sets.

Is it possible to solve the problems in a similar way as in the case of binary automata?
It is well-known that every p-valued function can be represented as a polynomial modulo p if p is a prime number. If p is not a prime number, then only a subset of p-valued functions may be represented by polynomials modulo p.

In the cases where F, G and g_t of an automaton $\mathfrak{A}$ are p-valued functions representable by polynomials, it should be possible to characterize (F,G)-superponable automata by solving the corresponding equations of polynomials. But there is one obstacle. We do not know any good classification of all p-valued functions. Therefore it seems that is has been impossible till now to formulate general results concerning superponability of automata with respect to arbitrary 2-ary p-valued pairs of functions F and G as it is possible in the binary case. To overcome these difficulties in the following we assume that some algebraic properties, like associativity, commutativity, bisymmetry,
are to be satisfied by F and G. Of course we do not have such general results as in the binary case.

3.3. (o,▽)-linear automata

Now we investigate (o,▽)-superponable automata if both the operations o and ▽ are associative and commutative. It will be shown that in this case (o,▽)-superponable automata are (o,▽)-linear.

First we introduce the concept of (o,▽)-linear functions and (o,▽)-linear automata.

Definition 4. Let o and ▽ be associative and commutative operations on X and Y, respectively, i. e.

$$(x \circ x) \circ x'' = x \circ (x \circ x'') \quad \text{for } x,x',x'' \in X \qquad \text{associativity}$$
$$(y \triangledown y') \triangledown y'' = y \triangledown (y' \triangledown y'') \quad \text{for } y,y',y'' \in Y$$

and

$$x \circ x' = x' \circ x \quad \text{for } x,x' \in X \qquad \text{commutativity}$$
$$y \triangledown y' = y' \triangledown y \quad \text{for } y,y' \in Y$$

Let $H_{o,\triangledown}$ be the set of all unary functions $h : X \rightarrow Y$ that satisfy

$$h(x \circ x') = h(x) \triangledown h(x') \tag{53}$$

i. e.

$$H_{o,\triangledown} = \{h/h(x \circ x') = h(x) \triangledown h(x'), \text{ if } x,x' \in X\} \tag{54}$$

and let $f_n : X^n \rightarrow Y$, $n > 0$, an n-ary function. Then f_n is called (o,▽)-linear if f_n can be represented as

$$f_n(x_1,\dots,x_n) = h_1(x_1) \triangledown \dots \triangledown h_n(x_n) \tag{55}$$

with $h_i \in H_{o,\triangledown}$, $i = 1,\dots,n$.

An n-ary (o,▽)-linear function is composed by unary (o,▽)-superponable functions.

Definition 5. Let $\mathfrak{a} = (X,Y,Z,z_o,\delta,\lambda)$. Then $\mathfrak{a}$ is called (o,▽)-linear if for all $t > 0$ functions $g_t : X^t \rightarrow Y$ are (o,▽)-linear.

It is easy to see that (o,▽)-linearity is a natural generalization of ordinary linearity. We show that (+,+)-linearity and ordinary linearity coincide. For simplicity let X = Y = K, where K is a finite or infinite field.

H_{++} is the set of all unary functions $h : X \rightarrow Y$ that satisfy

$$h(x + x') = h(x) + h(x') \tag{56}$$

Eq. (56) is a well-known functional equation with the solution

$$h(x) = cx, \qquad c \in K \tag{57}$$

(s. e. g. /1/). Thus an n-ary function $f_n : X^n \to Y$ is (+,+)-linear if it can be expressed as

$$f_n(x_1,\dots,x_n) = \sum_{i=1}^{n} c_i\, x_n \tag{58}$$

i. e. if f_n is linear.

The next theorem may be considered as one of the main theorems of this chapter. It shows the close connection between (o,∇)-superponability and (o,∇)-linearity /34, 13/.

Theorem 14. (Characterization theorem of (o,∇)-linear automata) Let $\mathfrak{A} = (X,Y,Z,z_0,\delta,\lambda)$ and let o and ∇ be associative and commutative operations and let e be the unity elements of o. Then $\mathfrak{A}$ is (o,∇)-superponable if and only if $\mathfrak{A}$ is (o,∇)-linear.

Proof.

a. Let $\mathfrak{A}$ be (o,∇)-linear. Then for all $t > 0$, $x_1,\dots,x_t, x'_1,\dots,x''_t \in X$ we have $g_t(x_1 \circ x'_1,\dots,x_t \circ x'_t) = h^t_1(x_1 \circ x'_1) \nabla \dots \nabla h^t_t(x_t \circ x'_t) =$

$= h^t_1(x_1) \nabla h^t_1(x'_1) \nabla \dots \nabla h^t_t(x_t) \nabla h^t_t(x'_t) =$

$= h^t_1(x_1) \nabla \dots \nabla h^t_t(x_t) \nabla h^t_1(x'_1) \nabla \dots \nabla h^t_t(x'_t) =$

$= g_t(x_1,\dots,x_t) \nabla g_t(x'_1,\dots,x'_t)$, with $h^t_i \in H_{o\nabla}$, $i = 1,\dots,t$.

b. Let $\mathfrak{A}$ be (o,∇)-superponable. Then for every $t > 0$ we define

$$h^t_i(x) = g^{\mathfrak{A}}_t(\underbrace{e,\dots,e}_{i-1}, x, e \dots e) \qquad i = 1,\dots,t. \tag{59}$$

Then we have $h^t_i(x \circ x') = g_t(\underbrace{e,\dots,e}_{i-1}, x \circ x', e,\dots,e) = g_t(e \circ e,\dots$

$\dots, e \circ e, x \circ x', e \circ e,\dots,e \circ e)$

$= g^{\mathfrak{A}}_t(\underbrace{e,\dots,e}_{i-1},x,e,\dots,e) \nabla g^{\mathfrak{A}}_t(\underbrace{e,\dots,e}_{i-1},x',e,\dots,e) = h^t_i(x) \nabla h^t_i(x')$,

i. e. $h^t_i \in H_{o,\nabla}$, $i = 1,\dots,t$ and

$h^t_1(x_1) \nabla \dots \nabla h^t_t(x_t) = g_t(x_1,e,\dots,e) \nabla \dots \nabla g_t(e,\dots,e,x_t) =$

$g_t(x_1 \circ e \circ \dots \circ e,\dots,e \circ \dots \circ e \circ x_t) = g_t(x_1,\dots,x_t)$. ■

Remark 15. If we want to determine the explicit form of the functions $g^{\mathfrak{A}}_t$ of an (o,∇)-superponable i. e. (o,∇)-linear automaton instead of the infinite system of functional equations

$$g_t^{\mathfrak{a}}(x_1 \circ x'_1, \ldots, x_t \circ x'_t) = g_t^{\mathfrak{a}}(x_1, \ldots, x_t) \triangledown g_t^{\mathfrak{a}}(x'_1, \ldots, x'_t) \quad t = 1,2,3,\ldots$$

by Theorem 14 we only have to solve the much simpler functional equation

$$h(x \circ x') = h(x) \triangledown h(x').$$

Now we discuss different examples of associative and commutative operations o and $\triangledown$ and corresponding (o,$\triangledown$)-linear automata.

1. Let $X = Y = \{0,1\}$, $x \circ x' = x \triangledown x' = x \vee x'$ ($\vee$: disjunction). The operation $\vee$ is associative and commutative, 0 is the unity element. Eq. (53) becomes

$$h(x \vee x') = h(x) \vee h(x'), \tag{60}$$

where h is a unary Boolean functions. $h_1(x) = x$, $h_2(x) = 1$, $h_3(x) = 0$ are the solutions of Eq. (60) and $g_t^{\mathfrak{a}}$ of the corresponding (o,$\triangledown$)-linear automaton $\mathfrak{a}$ is for $t > 0$

$$g_t(x_1, \ldots, x_t) = c_o^t \bigvee_{i \in I(t)} x_i \tag{61}$$

where $I(t) \subseteq \{1, \ldots, t\}$ is an index set in accordance with Theorem 9. ($c_j^t = 1$ iff $j \in I(t)$).

2. Let $X = Y = \mathbb{R}$ (real numbers) and let o and $\triangledown$ be rational operations

$$x \circ x' = \frac{P(x,x')}{Q(x,x')} \qquad y \triangledown y' = \frac{R(y,y')}{S(y,y')} \tag{62}$$

where $P(x,x')$, $Q(x,x')$, $R(y,y')$, $S(y,y')$ are polynomials of finite degree in x,x' and y,y', respectively, /41/.

Furthermore we assume:

1. The pairs of polynomials P,Q and R,S are relatively prime
2. Neither o nor $\triangledown$ are identically constant or infinite.

Then the following theorem can be proved /29/:

<u>Theorem 15.</u> Let o and $\triangledown$ be associative, commutative and rational operations for which the assumptions 1. and 2. are fulfilled, then o and $\triangledown$ are of the form

$$x \circ x' = \frac{l\,x\,x' + m\,x + mx' + n}{p\,x\,x' + q\,x + q\,x' + r} \tag{63}$$

$$y \nabla y' = \frac{L\,y\,y' + M\,y + M\,y' + N}{P\,y\,y' + Q\,y + Q\,y' + R} \qquad (63)$$

where l, m, n, p, q, r, L, M, N, P, Q, R $\in \mathbb{R}$ are real constants, for which the conditions

$$\text{rank}\begin{pmatrix} m & l - q & p \\ n & m - r & q \end{pmatrix} = 1 \qquad \text{rank}\begin{pmatrix} M & L - Q & P \\ N & M - R & Q \end{pmatrix} = 1 \qquad (64)$$

$l : m : n \neq p : q : r$ and $L : M : N \neq P : Q : Q$ hold.
The unity element e_o of the operation o is uniquely determined by

$$m\,e_o + n = p\,e_o + q = (1 - q)\,e_o + m - r = 0 \qquad (65)$$

in accordance with (64).
The concrete form of Eq. (60) is

$$h\left(\frac{l\,x\,x' + m\,x + mx' + n}{p\,x\,x' + q\,x + q\,x' + r}\right) = \qquad (66)$$

$$= \frac{L\,h(x)\,h(x') + M\,h(x) + Mh(x') + N}{P\,h(x)\,h(x') + Q\,h(x) + Q\,h(x) \pm R}$$

with the conditions (64).
For the case $l = p = q = n = 0$ and $m = r = 1$ the solutions of Eq. (66) are in /29/, and for $L = P = Q = 0$ and $M = R = 1$ considerations on the solutions of Eq. (66) are in /2/.

Some theorems of spherical trigonometry such as

$$h(x + y) = \frac{h(x) + h(y)}{1 - h(x)\,h(y)} \qquad (67)$$

with the solution $h(x) = tg(cx)$, $c \in \mathbb{R}$ are included in Eq. (66). ($l = n = p = q = 0$, $m = r = 1$, $L = N = Q = 0$, $P = 1$, $M = R = 1$).
Let us suppose h to be differentiable. Then the functional equation (53) can be reduced to an ordinary differential equation. Differentiating Eq. (66) with respect to x', setting thereafter $x' = e_o$, we obtain

$$\frac{dh}{dx} = c\,\frac{-P\,h^2 + (L - Q)\,h + M}{-p\,x^2 + (l - q)\,x + m} \qquad (68)$$

where

$$c = \left.\frac{dh}{dx}\right|_{e_o} \frac{q\,e_o + r}{Q\,e_\nabla + R},$$

with the condition $h(e_o) = e_\nabla$, where e_∇ is the unity element of the operation ∇. Neither $q\,e_o + r$ nor $Q\,e_\nabla + R$ can be zero, since

$x \circ e_o = x$ and $qe + r = 0$ or $y \nabla e_\nabla = y$ and $Qe_\nabla + R = 0$ imply

$$\frac{l\, e_o + m}{p\, e_o + q} + \frac{m\, e_o + n}{p\, e_o + q}\, \frac{1}{x} = x \quad \text{and} \quad \frac{L\, e_\nabla + M}{P\, e_\nabla + Q} + \frac{M\, e_\nabla + N}{P\, e_\nabla + Q}\, \frac{1}{y} = y,$$

which can not be satisfied for any choice of l, m, n, p, q and L, M, N, P, Q.

In principle the integration of Eq. (68) can be done without difficulties by separation and partial fraction expansion ect., but many different cases are to be considered.

As a simple example let us deal here with polynomial operations

$$\begin{aligned} x \circ x' &= lxx' + m\,x + m\,x' + n \quad \text{and} \\ y \nabla y' &= Lyy' + M\,y + M\,y' + N \end{aligned} \tag{69}$$

with

$$\operatorname{rank}\begin{pmatrix} m & l & 0 \\ n & m-1 & 0 \end{pmatrix} = 1 \qquad \operatorname{rank}\begin{pmatrix} M & L & 0 \\ N & M-1 & 0 \end{pmatrix} = 1, \tag{70}$$

$$n = \frac{m^2 - m}{l}, \quad N = \frac{M^2 - M}{L}, \quad \text{if } l, L \neq 0,$$

and

$$e_o = \frac{1 - m}{l}, \quad e_\nabla = \frac{1 - M}{L}.$$

Then Eq. (53) becomes

$$\begin{aligned} &h(lxx' + mx + mx' + \frac{m^2 - m}{l}) = \\ &= Lh(x)\, h(x') + Mh(x) + Mh(x') + \frac{M^2 - M}{L}. \end{aligned} \tag{71}$$

Differentiating Eq. (71) with respect to x' we obtain

$$h'(lxx' + mx + mx' + \frac{m^2 - m}{l}) \{ l\, x + m \} = L\, h(x)\, h'(x') + M\, h'(x'). \tag{72}$$

By setting $x' = \frac{1 - m}{l} = e_o$ Eq. (72) reduces to

$$h'(x) = c\, \frac{Lh(x) + M}{l\, x + m} \tag{73}$$

where

$$c = h'\left(\frac{1-m}{l}\right)$$

in accordance with Eq. (68).
Written in differential form Eq. (73) becomes

$$\frac{dh}{Lh + M} = c\,\frac{dx}{lx + m} \tag{74}$$

and the general solution of Eq. (74) is

$$h_c(x) = \frac{1}{L}\,(lx + m)^{\frac{L}{l}c} - \frac{M}{L}\,. \tag{75}$$

where c,c $\in \mathbb{R}$ is the free parameter.
For $t > 0$ the function $g_t^{\mathfrak{a}}$ of the corresponding (o,∇)-linear automaton $\mathfrak{a}$ is

$$g_t^{\mathfrak{a}}(x_1,\dots,x_t) = \frac{1}{L}\left\{(lx_1 + m)^{\frac{L}{l}c_1^t} - M\right\} \nabla \dots \dots \nabla \frac{1}{L}\left\{(lx_t + m)^{\frac{L}{l}c_t^t} - M\right\} \tag{76}$$

In /41/ Eq. (68) is discussed in detail and the general solution for all possible parameter constellations is given.

3. Quasiadditions. Now we suppose the operations o and ∇ to be quasiadditions.

<u>Definition 6.</u> Let $X = \mathbb{R}$ (real numbers). Then the operation o on X is called quasiaddition iff for all $x,x' \in X$, $x \circ x'$ can be expressed as

$$x \circ x' = \varphi(\varphi^{-1}(x) + \varphi^{-1}(x')) \tag{77}$$

where φ is a strictly monotonic (i. e. invertible) function and φ^{-1} denotes the inverse function of φ.

<u>Definition 7.</u> Let $X = E_p = \{0,1,\dots,p-1\}$ and p a prime number. Then the operation o on X is called quasiaddition iff for all $x,x' \in X$, $x \circ x'$ can be expressed as

$$x \circ x' = \pi(\pi^{-1}(x) + \pi^{-1}(x')) \tag{78}$$

where $\pi : X \rightarrow X$ is a permutation, π^{-1} is the inverse permutation of π and + denotes addition modulo p.

It is of interest to notice here that associative resolvable operations are quasiadditions.

Theorem 16. Let $F(x,x') = x \circ x'$, $x,x' \in X = \mathbb{R}$ (real numbers).

If F is associative, continuous and resolvable with respect to both of its variables, i. e. for $x,x',u \in X$

$$F(x,u) = F(x',u) \quad \text{implies} \quad x = x'$$
$$F(u,x) = F(u,x') \quad \text{implies} \quad x = x',$$

then $F(x,x')$ is a quasiaddition.

The proof is in /1/.

Theorem 16'. Let $F(x,x') = x \circ x'$; $x,x' \in X = E_p$ and let p be a prime number.
If F is associative and resolvable with respect to both its variables, then F is a quasiaddition.

Proof. Since $F(x,x')$ is associative, F is bisymmetric. Therefore, for all $x,x',z,z' \in X$ we have

$$F(F(x,x'), F(z,z')) = F(F(x,z'), F(z,x')).$$

In /18/ it is proved that every bisymmetric resolvable function $F(x,x')$ can be expressed as

$$F(x,x') = \pi(a\pi^{-1}(x) + b\pi^{-1}(x') + c), \quad a,b,c \in E_p \tag{80}$$

Since F is associative, we conclude $a = b = 1$, $c = 0$. ■

If $x \circ x' = F(x,x')$ is a quasiaddition, then it is easy to determine the function φ in Eq. (79) if φ is differentiable.
Eq. (79) is equivalent to

$$\varphi^{-1}(F(x,x')) = \varphi^{-1}(x) + \varphi^{-1}(x') \tag{81}$$

Differentiating Eq. (81) with respect to x' and putting thereafter $x' = e = \varphi(o)$ we obtain

$$\varphi^{-1\prime}(x)\, F_{x'}(x,x')\Big|_{x'=e} = \varphi^{-1\prime}(e) = c$$

$$\varphi^{-1}(x) = \int^{x} \frac{c\, dx}{F_{x'}(x,x')\Big|_{x'=e}} \tag{82}$$

with the condition $\varphi^{-1}(e) = 0$.
Since the operations o and ∇ are quasiadditions, we are in the position to derive a canonical representation of (o,∇)-superponable automata. It is possible to show that every (o,∇)-superponable automaton is

equivalent to a serial connection of the combinational function φ^{-1}, a linear automaton $\mathcal{L}$ and the combinational function Ψ. The functions φ^{-1} and Ψ are determined by the quasiadditions $x \circ x' = \varphi(\varphi^{-1}(x) + \varphi^{-1}(x')$ and $y \nabla y' = \Psi(\Psi^{-1}(y) + \Psi^{-1}(y')$. Thus different $(\circ,\nabla)$-superponable automata differ only in their corresponding linear automata. These results are very similar to those obtained in /31/ for so-called **homomorphic systems** (s. chapter 5).

Theorem 17. Let $x \circ x' = \varphi(\varphi^{-1}(x) + \varphi^{-1}(x'))$ for $x,x' \in X$ and $y \nabla y' = \Psi(\Psi^{-1}(y) + \Psi^{-1}(y'))$ for $y,y' \in Y$. Then the automaton $\mathfrak{A} = (X,Y,Z,z_o,\delta,\lambda)$ is $(\circ,\nabla)$-superponable if and only if there exists a linear automaton $\mathcal{L}_{\mathfrak{A}} = (X,Y,Z_{\mathcal{L}},z_{o\mathcal{L}},\delta_{\mathcal{L}},\lambda_{\mathcal{L}})$ such that the serial connection $\varphi^{-1} - \mathcal{L}_{\mathfrak{A}} - \Psi$ is equivalent to $\mathfrak{A}$.

Proof. a) (If $\varphi^{-1} - \mathcal{L}_{\mathfrak{A}} - \Psi$ is equivalent to $\mathfrak{A}$, then $\mathfrak{A}$ is $(\circ,\nabla)$-superponable.)
Let

$$g_t^{\mathcal{L}_{\mathfrak{A}}}(x_1,\ldots,x_t) = \sum_{i=1}^{t} c_i^t\, x_i, \qquad t > 0.$$

Hence

$$g_t^{\mathfrak{A}}(x_1,\ldots,x_t) = \Psi(\sum_{i=1}^{t} c_i^t \varphi^{-1}(x_i)),$$

and therefore

$$g_t^{\mathfrak{A}}(x_1 \circ x_1',\ldots,x_t \circ x_t') = \Psi\{\sum_{i=1}^{t} c_i^t(\varphi^{-1}(x_i) + \varphi^{-1}(x_i'))\} =$$

$$= \Psi\{\sum c_i^t \varphi^{-1}(x_i) + c_i^t \varphi^{-1}(x_i')\} =$$

$$= g_t^{\mathfrak{A}}(x_1,\ldots,x_t) \nabla g_t^{\mathfrak{A}}(x_1',\ldots,x_t').$$

b) (If $\mathfrak{A}$ is $(\circ,\nabla)$-superponable, then $\mathfrak{A}$ is equivalent to $\varphi^{-1} - \mathcal{L}_{\mathfrak{A}} - \Psi$)
Let $\mathfrak{A}$ be $(\circ,\nabla)$-superponable, then we have

$$g_t^{\mathfrak{A}}(x_1,\ldots,x_t) = h_1^t(x_1) \nabla \ldots \nabla h_t^t(x_t) \tag{83}$$

with

$$h_i^t(\varphi(\varphi^{-1}(x) + \varphi^{-1}(x') = \Psi\{\Psi^{-1}(h_i^t(x)) + \Psi^{-1}(h_i^t(x'))\}$$

or

$$\Psi^{-1}\{h_i^t(\varphi(\varphi^{-1}(x) + \varphi^{-1}(x'))\} = \Psi^{-1}(h_i^t(x)) + \Psi^{-1}(h_i^t(x')) \tag{84}$$

Substituting $x = \varphi(u)$, $x' = \varphi(v)$ Eq. (84) becomes

$$\psi^{-1}(h_i^t(\varphi(u+v)) = \psi^{-1}\{h_i^t(\varphi(u))\} + \psi^{-1}\{h_i^t(\varphi(v))\} \tag{85}$$

with the solution

$$\psi^{-1}(h_i^t(\varphi(x))) = c_i^{t'}x. \tag{86}$$

Hence it follows that

$$g_t^{\mathfrak{a}}(x_1,\ldots,x_t) = \sum_{i=1}^{t} c_i^t\,\varphi^{-1}(x_i) \tag{87}$$

and $\mathfrak{a}$ is equivalent to $\varphi^{-1} - \mathcal{L}_{\mathfrak{a}} - \psi$ with

$$g_t^{\mathcal{L}_{\mathfrak{a}}}(x_1,\ldots,x_t) = \sum_{i=1}^{t} c_i^t\,x_i. \blacksquare \tag{88}$$

The serial connection $\varphi^{-1} - \mathcal{L}_{\mathfrak{a}} - \psi$ equivalent to $\mathfrak{a}$ is called canonical representation of $\mathfrak{a}$. The canonical representation of an automaton $\mathfrak{a}$ (o,∇)-superponable with respect to quasiadditions $x \circ x' = \varphi(\varphi^{-1}(x) + \varphi^{-1}(x'))$ and $y \nabla y' = \psi(\psi^{-1}(y) + \psi^{-1}(y'))$ is represented in Fig. 8.

Fig.8 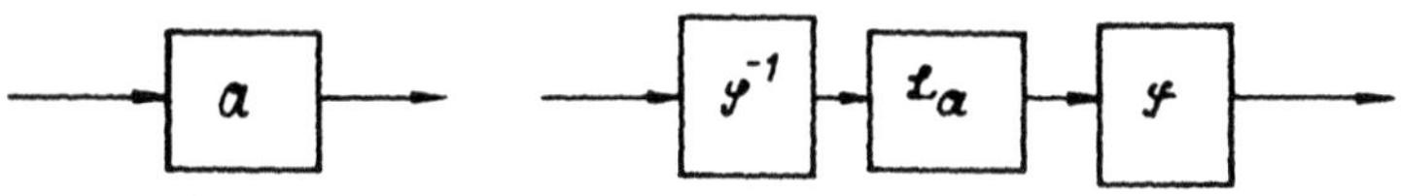

Automata superponable with respect to quasiadditions are invertible.

<u>Theorem 18.</u> Let $\mathfrak{a} = (X,Y,Z,z_0,\delta,\lambda)$ be (o,∇)-superponable and let $x \circ x' = \varphi(\varphi^{-1}(x) + \varphi^{-1}(x'))$, $y \nabla y' = \psi(\psi^{-1}(y) + \psi^{-1}(y'))$ be quasiadditions. Then $\mathfrak{a}$ is invertible.

Proof. Since $\mathfrak{a}$ is (o,∇)-superponable, we have for $t > 0$

$$g_t^{\mathfrak{a}}(x_1,\ldots,x_t) = h_1^t(x_1) \nabla \ldots \nabla h_t^t(x_t)$$

and by Eq. (86)

$$h_i^t(x_i) = \psi(c_i^t\,\varphi^{-1}(x_i)). \tag{89}$$

For a given output sequence $y_1 \ldots y_t$ the corresponding input sequence $x_1 \ldots x_t$ can be uniquely determined step by step by

$$y_1 = h_1^1(x_1) = \psi(c_1^1\,\varphi^{-1}(x_1))$$

$$y_2 = h_1^2(x_1) \nabla h_2^2(x_2) = \Psi(c_1^2 \varphi^{-1}(x_1) + c_2^2 \varphi^{-1}(x_2))$$

$$\vdots$$

$$y_t = h_1^t(x_1) \nabla \dots \nabla h_t^t(x_t) = \Psi(\sum_{i=1}^{t} c_i^t \varphi^{-1}(x_i)). \blacksquare$$

As an example let us consider $X = Y = \mathbb{R}$,

$$x \circ x' = Lxx' + Mx + mx' + \frac{M^2 - M}{L} = \varphi(\varphi^{-1}(x) + \varphi^{-1}(x')) \qquad (91)$$

$$y \nabla y' = \frac{y + y'}{1 - yy'} = \Psi(\Psi^{-1}(y) + \Psi^{-1}(y')) . \qquad (92)$$

By use of Eq. (82) we obtain

$$\varphi^{-1}(x) = \int^{x} \frac{c_1 \, dx}{Lx + M} = \frac{c_1}{L} \log(Lx + M) \qquad (93)$$

$$\Psi^{-1}(y) = \int^{y} \frac{c_2 \, dy}{1 + y^2} = c_2 \operatorname{arctg}(y) \qquad (94)$$

and therefore

$$\varphi(x) = \frac{1}{L} \{\exp. (\frac{Lx}{c_1}) - M\}, \qquad (95)$$

$$\Psi(y) = \operatorname{tg} \frac{y}{c_2} . \qquad (96)$$

If the automaton $\mathfrak{A}$ is $(\circ,\nabla)$-superponable, then by Theorem 17, Eq. (87), the function $g_t^{\mathfrak{A}}$ can be expressed as

$$g_t^{\mathfrak{A}}(x_1,\dots,x_t) = \operatorname{tg} \{ \sum_{i=1}^{t} \tilde{c}_i^t \log (Lx_i + M)\} \qquad (97)$$

with

$$\tilde{c}_i^t = \frac{c_i^t \, c_1}{c_2 \, L}$$

and different automata superponable with respect to the quasiadditions $\circ$ and ∇ defined by Eqs. (91) and (92), respectively, differ only by their coefficients c_i^t in Eq. (97), i. e. by their linear part.

Sometimes, for instance in control theory (cf. e. g. /7/) a system S is considered as a serial connection of a linear dynamic system L_1 and one or two nonlinear static systems N_1 and N_2. If N_1 and N_2 are invertible, S is superponable with respect to a quasiaddition.

If the static system is nonlinear, sometimes its nonlinear characteristic can be approximated by an invertible characteristic and S is approximately superponable with respect to a quasiaddition. Thus the function

$$s(x) = \begin{cases} 0 & \text{if } x \leq 0 \\ 1 & \text{if } x > 0 \end{cases}$$

can be approximated by the invertible function

$$s_\varepsilon(x) = \frac{1}{2} + \frac{1}{\pi} \text{ arc tg } \frac{x}{\varepsilon}$$

$$0 < \varepsilon \ll 1.$$

Now we ask under which conditions the input-output-behaviour of (o,∇)-linear automata can be described by generalized impulse responses. The answer to this question is given by the following theorems.

Theorem 19. Let $\mathfrak{A} = (X,Y,Z,z_o,\delta,\lambda)$ be a (o,∇)-linear automaton and let e be the unity element of the operation o. Then for every $t > 0$, $x_1,\ldots,x_t \in X$, the output word $\lambda(x_1\ldots x_t)$ is uniquely determined by $\lambda(x_1e\ldots e)$, $\lambda(ex_2e\ldots e),\ldots, \lambda(e\ldots ex_t)$.

Proof. By Theorem 14 $\mathfrak{A}$ is (o,∇)-superponable. Hence

$\lambda(x_1\ldots x_t) = \lambda(x_1e\ldots e) \nabla \lambda(ex_2e\ldots e) \nabla \ldots \nabla \lambda(e\ldots ex_t)$

If X is a finite set with $|X| = n$, the n^t possible words of the lenght t are determined by $n\cdot t$ output-words. ■

Theorem 20. Let $\mathfrak{A} = (X,Y,Z,z_o,\delta,\lambda)$ be a (o,∇)-linear automaton and let $h(e) = E$ if $h \in H_{o,\nabla}$, where e and E are the unity elements of o and ∇, respectively. If there exists an $x_o \in X$ such that for all h, $k \in H_{o,\nabla}$ $h(x_o) = k(x_o)$ implies $h = k$, then for $t > 0$, $x_1,\ldots,x_t \in X$ the output word $\lambda(x_1\ldots x_t)$ is determined by the t output words

$\lambda(x_o\underbrace{e\ldots e}_{t})$, $\lambda(ex_oe\ldots e),\ldots, \lambda(e\ldots ex_o)$.

Proof. Let $\lambda(x_1\ldots x_t) = y_1\ldots y_t$. Since $\mathfrak{A}$ is (o,∇)-linear, we have $y_1 = h_1^1(x_1)$, $y_2 = h_1^2(x_1) \nabla h_2^2(x_2),\ldots,y_t = h_1^t(x_1) \nabla \ldots \nabla h_t^t(x_t)$ and $\lambda(x_1\ldots x_t)$ is determined if $h_1^1, h_1^2, h_2^2,\ldots,h_t^t$ are known. On the other hand we have

$\lambda(x_oe\ldots e) = h_1^1(x_o)\ h_1^2(x_o)\ldots h_1^t(x_o)$

$\lambda(ex_oe\ldots e) = E\ h_2^2(x_o)\ldots h_2^t(x_o)$

$\vdots \qquad \vdots \quad \vdots \qquad \vdots$

$$\begin{array}{cccc} \vdots & \vdots & \vdots & \vdots \\ \lambda(e\ldots ex_o) = E & E & \ldots & h_t^t(x_o) \end{array}$$

For $h \in H_{o,\nabla}$ the function h is determined by $h(x_o)$. ■

If the operations o and ∇ are quasiadditions, the conditions of Theorem 20 are satisfied.

<u>Theorem 21.</u> Let $x \circ x' = \varphi(\varphi^{-1}(x) + \varphi^{-1}(x'))$, $x,x' \in X$ and $y \nabla y' = \psi(\psi^{-1}(y) + \psi^{-1}(y'))$, $y,y' \in Y$ be quasiadditions.
Then for all $h_1,h_2 \in H_{o\nabla}$ $h_1(x_o) = h_2(x_o)$ implies $h_1 = h_2$.

Proof. The solution of the equation

$$h(\varphi(\varphi^{-1}(x) + \varphi^{-1}(x')) = \psi(\psi^{-1}(h(x)) + \psi^{-1}(h(x'))$$

is $h(x) = \psi(c\,\varphi^{-1}(x))$, $\quad c = \text{const.}$

(s. proof of Theorem (17)). Therefore h_1, h_2 can be expressed as

$$h_1(x) = \psi(c_1\varphi^{-1}(x)),\ h_2(x) = \psi(c_2\varphi^{-1}(x)).$$

From $h_1(x_o) = h_2(x_o)$ we conclude $c_1 = c_2$, i. e. $h_1 = h_2$. ■

<u>Theorem 22.</u> Let $\mathfrak{A} = (X,Y,Z,z_o,\delta,\lambda)$ be a (o,∇)-superponable automaton, let $h(e) = E$ if $h \in H_{o,\nabla}$, where e and E are the unity elements of o and ∇, respectively, and let $\delta(z_o,e) \sim z_o$.
Then the input-output-behaviour of $\mathfrak{A}$ is completely determined by the generalized impulse responses $\lambda(x\ e\ e\ e\ \ldots)$, $x \in X$.

Proof. Since $\mathfrak{A}$ is (o,∇)-superponable, $\mathfrak{A}$ is (o,∇)-linear and we conclude $\lambda(e\ e\ e\ \ldots) = E\ E\ E\ \ldots$ and for $k > 0$, $x \in X$
$\lambda(e\ldots e\ x\ eee\ \ldots) = E\ \ldots\ E\ \lambda(x\ eee\ \ldots)$. For every $t > 0$, $x_1,\ldots,x_t \in X$ the output word $\lambda(x_1\ldots x_t)$ is determined by $\lambda(x_1e\ldots e)$, $\lambda(ex_2e\ldots \ldots e),\ldots,\ \lambda(e\ldots ex_t)$ by Theorem 19. ■

In many cases we do not need the knowledge of all generalized impulse responses $\lambda(xeee\ldots)$, $x \in X$. From known impulse responses $\lambda(x_1eee\ldots)$, $\lambda(x_2eee\ldots)$ we can compute $\lambda(x_1 \circ x_2,eee\ldots) = \lambda(x_1eee\ldots) \nabla \lambda(x_2eee\ldots)$ etc.
If $x \circ x' = \varphi(\varphi^{-1}(x) + \varphi^{-1}(x'))$ and $y \nabla y' = \psi(\psi^{-1}(y) + \psi^{-1}(y'))$ are quasiadditions, then every $\lambda(x_oeee\ldots)$, $x_o \in X$, determines for all $x \in X$ the generalized impulse responses $\lambda(xeee\ldots)$.

<u>Theorem 23.</u> Let $\mathfrak{A}$ be a (o,∇)-superponable automaton where $x \text{ o } x' = \varphi(\varphi^{-1}(x) + \varphi^{-1}(x'))$ and $y \nabla y' = \psi(\psi^{-1}(y) + \psi^{-1}(y))$ are quasiadditions and let $\delta(z_o,e) \sim z_o$, $e = \varphi(0)$. Then for every $x_o \in X$ the input-output-behaviour of $\mathfrak{A}$ is completely determined by $\lambda(x_o eee..)$.

Proof. Theorem 23 is a direct consequence of the Theorems 19, 20 and 21. As in the case of linear automata, $\mathfrak{A}$ can be identified by $\lambda(x_o eee...)$. If we liked to model $\mathfrak{A}$ by means of $\lambda(x_o eee...)$, we could do this in the form of its canonical representation $\varphi^{-1} - \mathcal{L}_{\mathfrak{A}} - \psi$ (cf. Theorem 17 and Fig. 8).

Let $\lambda(x_o eee...) = y_1\, y_2\, y_3 \,...$ If the input sequence of $\varphi^{-1} - \mathcal{L}_{\mathfrak{A}} - \psi$ is $x_o eee...$, then the input sequence of $\mathcal{L}_{\mathfrak{A}}$ is $\varphi^{-1}(x_o)000...$ and the corresponding output sequence of $\mathcal{L}_{\mathfrak{A}}$ is

$\{\varphi^{-1}(x_o)\; H(0)\}\; \{\varphi^{-1}(x_o)\; H(1)\}\; \{\varphi^{-1}(x_o)\; H(2)\}\; ...$

Thus we obtain

$$H(0) = \frac{\psi^{-1}(y_1)}{\varphi^{-1}(x_o)}\,, \quad H(1) = \frac{\psi^{-1}(y_2)}{\varphi^{-1}(x_1)}\,, \quad H(2) = \frac{\psi^{-1}(y_3)}{\varphi^{-1}(x_2)}\,, \; ...$$

and the impulse response $H(t)$ of $\mathcal{L}_{\mathfrak{A}}$ is determined by x_o and by $\lambda(x_o eee...)$. Therefore the automaton $\mathcal{L}_H$ with the same input-output-behaviour as $\mathcal{L}_{\mathfrak{A}}$ can be realized as described in chapter 2 and $\varphi^{-1} - \mathcal{L} - \psi$ is equivalent to $\mathfrak{A}$. ■

The purpose of the Theorems 10 to 23 was to show how to generalize the description of the input-output-behaviour by means of generalized impulse responses well-known in the theory of linear automata to (o,∇)-linear automata.

Till now we have considered associative and commutative operations o and ∇ only. As we know (cf. Theorem 14), in the case of associative and commutative operations (o,∇)-superponability and (o,∇)-linearity coincide. Moreover, for $n > 0$ every n-ary (o,∇)-linear function $f_n(x_1,...,x_n)$ can be expressed as a composition of unary (o,∇)-superponable functions

$$f_n(x_1,...,x_n) = h_1(x_1) \nabla \;...\; \nabla\, h_n(x_n) \qquad (55)$$

with

$$h(x \text{ o } x') = h(x) \nabla h(x').$$

A similar result is in operation under some weaker conditions. To show this let us consider again the set $H_{o,\nabla}$ of unary (o,∇)-superponable functions $h : X \rightarrow Y$

$$H_{o,\nabla} = \{h \;/\; h(x \circ x') = h(x) \nabla h(x'),\quad x,x' \in X\}. \tag{98}$$

Furthermore let us introduce a third 2-ary operation $\square$ on Y. With this operation we define an n-ary function $f_n \; X^n \to Y$ as a composition of unary functions from $H_{o,\nabla}$.

Definition 8. The function

$$f_n(x_1,\ldots,x_n) = (\ldots(h_1(x_1) \square h_2(x_2)) \square \ldots) \square h_n(x_n) \tag{99}$$

with $h_1,\ldots,h_n \in H_{c,\nabla}$

is called $(o,\nabla,\square)$-linear.

If $y \nabla y' = y \square y'$ for all $y,y' \in Y$ and o and ∇ associative, f_n defined by Eq. (59) is (o,∇)-linear. In general $o, \nabla, \square$ are not supposed to be associative. Similar to the concept of (o,∇)-linear automata we introduce the concept of $(o, \nabla, \square))$-linear automata.

Definition 9. Let $\mathfrak{A} = (X,Y,Z,z_o,\delta,\lambda)$ be an automaton, then $\mathfrak{A}$ is called $(o,\nabla,\square)$-linear if for $t > 0$ g_t can be expressed as

$$g_t^{\mathfrak{A}}(x_1,\ldots,x_t) = (\ldots(h_1^t(x_1) \square h_2^t(x_2)) \square \ldots) \square h_t^t(x_t) \tag{100}$$

with $h_i^t \in H_{o,\nabla}$, $i = 1,\ldots,t$.

Now we deal with the question under which conditions $(o,\nabla,\square)$-linear automata are (o,∇)-superponable. An answer is given in the following theorem.

Theorem 24. Let $\mathfrak{A} = (X,Y,Z,z_o,\delta,\lambda)$ be a $(o,\nabla,\square)$-linear automaton. Then $\mathfrak{A}$ is (o,∇)-superponable if we have for all $y,u,v,w \in Y$

$$(y \nabla u) \square (v \nabla w) = (y \square v) \nabla u \square w). \tag{101}$$

Proof. Since $\mathfrak{A}$ is $(o,\nabla,\square)$-superponable, we have
$g_t^{\mathfrak{A}}(x_1 \circ x_1',\ldots,x_t \circ x_t') = (\ldots(h_1^t(x_1) \nabla h_1^t(x_1')) \square (h_2^t(x_2) \nabla h_2^t(x_2')) \square$
$\square \ldots) \square (h_t^t(x_t) \nabla h_t^t(x_t'))$ and by use of Eq. (101)
$g_t^{\mathfrak{A}}(x_1 \circ x_1',\ldots,x_t \circ x_t') = g_t^{\mathfrak{A}}(x_1,\ldots,x_t) \nabla g_t^{\mathfrak{A}}(x_1',\ldots,x_t')$. ■

By setting $y \nabla y' = F(y,y')$, $y \square y' = G(y,y')$ Eq. (101) becomes

$$G(F(y,u),\, F(v,w)) = F(G(y,v),\, G(u,w)) \tag{102}$$

Eq. (102) is a generalized equation of bisymmetry (s. e. g. /1/).
If F = G, Eq. (102) reduces to the ordinary equation of bisymmetry

$$F(F(y,u),\, F(v,w)) = F(F(u,v),\, F(u,w)) \tag{103}$$

To solve the functional equations (102) or (103) is an independent problem. We present here some known results and some simple examples only.

1. If $F(y,y')$ is a continuous solution of Eq. (103) resolvable with respect to x and y, then it can be represented as

$$F(y,y) = \varphi(a\, \varphi^{-1}(y) + b\, \varphi^{-1}(y') + c),$$

where φ is a strictly monotonic function, a, b, c are constants. If $a = b = 1$, $c = 0$, $F(x,y)$ is a quasiaddition and associative. For $X = \{0,1,\ldots,p-1\}$, p a prime number, we have a similar result (s. proof of Theorem 16').

2. If $x \circ x'$ is associative and commutative, then $x \,\square\, x' = x \circ x \circ y \circ y$ satisfies the equation of bisymmetry Eq. (103). For example let

$$x \circ x' = Lxx' + Mx + Mx' + \frac{M^2 - M}{L} \tag{104}$$

Then $x \circ x'$ is associative and commutative, (L, M constant). Therefore

$$x \,\square\, y = x \circ x \circ y \circ y = L^3x^2y^2 + 2ML^2(xy^2 + yx^2) + LM^2(x^2 + y^2) + + 4M^2Lxy + 2M^3(x + y) + \frac{M^4 - M}{L} \tag{105}$$

satisfies Eq. (103). But the polynomial defined by Eq. (105) is not associative.

3. Let $F(y,y') = \varphi(a\, \varphi^{-1}(y) + b\, \varphi^{-1}(y') + c)$,
$G(y,y') = \varphi(\alpha\, \varphi^{-1}(y) + \beta\, \varphi^{-1}(y') + \gamma)$,
where φ is a strictly monotonic function. a, b, c, α, β, γ are real constants.
Then F and G are solutions of Eq. (102) if

$$c(1 - \alpha - \beta) = \gamma(1 - a - b). \tag{106}$$

At the end of this chapter let us consider an example of a $(\circ, \nabla, \square)$-linear automaton. Let

$$\begin{aligned} x \circ x' &= \Psi(A\, \Psi^{-1}(x) + B\, \Psi^{-1}(x') + C), \\ y \,\nabla\, y' &= \varphi(a\, \varphi^{-1}(y) + b\, \varphi^{-1}(y') + c), \end{aligned} \tag{107}$$

where φ and Ψ are strictly monotonic functions and A, B, C, a, b, c are real constants.

The solution $h : X \to Y$ of the equation

$$h(x \circ x') = h(x) \nabla h(x') \tag{108}$$

is supposed to be of the form

$$h(x) = \varphi(d\, \psi^{-1}(x) + e) \tag{109}$$

where d and e are constants. Then

$$\begin{aligned} a &= A \\ b &= B \\ e(1 - A - B) &= c - dC \end{aligned} \tag{110}$$

have to be satisfied, and we have

$$\begin{aligned} x \circ x' &= \psi(a\, \psi^{-1}(x) + b\, \psi^{-1}(x') + c), \\ y \nabla y' &= \varphi(a\, \varphi^{-1}(y) + b\, \varphi^{-1}(y') + C) \end{aligned} \tag{111}$$

and

$$\begin{aligned} H_{o,\nabla} = \{h \,/\, h(x) &= \psi(d\, \varphi^{-1}(x) + e),\ e(1 - A - B) = \\ &= c - dC,\ d \in \mathbb{R}\} . \end{aligned} \tag{112}$$

A possible operation $\square$ is

$$y \square y' = \varphi(\alpha\, \varphi^{-1}(y) + \beta\, \varphi^{-1}(y') + \gamma) \tag{113}$$

with the condition described by Eq. (106), and the $(o, \nabla, \square)$-linear automaton $\mathfrak{a}$ with

$$\begin{aligned} g_t^{\mathfrak{a}}(x_1,\dots,x_t) = (\dots(\psi\,[\,d_1^t\, \varphi^{-1}(x_1) + e_1^t] \square \psi\,[\,d_2^t\, \varphi^{-1}(x_2) + \\ + c_2^t\,) \square \dots \square (\psi\,[\,d_t^t\, \varphi^{-1}(x_t) + c_t^t]) \end{aligned}$$

where

$$e_i^t(1 - A - B) = c - d_i^t C, \quad i = 1,\dots,t, \tag{114}$$

is (o,∇)-superponable.

4. Automata superponable with respect to pairs of automata

In this chapter we investigate the superposition of an automaton $\mathcal{A}$ with respect to a pair of automata $\mathcal{C}$ and $\mathcal{D}$. The 2-ary operations o and ∇ (without memory) of the previous chapter are replaced by automata $\mathcal{C}$ and $\mathcal{D}$ (with memory) with two input terminals and one output terminal each. It is shown that every non-autonomous automaton $\mathcal{A}$ is non-trivially $(\mathcal{C},\mathcal{D})$-superponable for a suitable choice of $\mathcal{C}$ and $\mathcal{D}$.

This result allows us to classify (nonlinear) automata by their superposition properties and to develop a uniform theory of nonlinear systems as a theory of nonlinear superposition.
If the automaton $\mathcal{A}$ is invertible, then it is possible to characterize all pairs of automata $(\mathcal{C},\mathcal{D})$ with respect to which $\mathcal{A}$ is $(\mathcal{C},\mathcal{D})$-superponable.
Some examples of $(\mathcal{C},\mathcal{D})$-superponable automata are considered.

As we have mentioned, till now we have investigated the superponability of an automaton $\mathcal{A} = (X,Y,Z,z_o,\delta,\lambda)$ with respect to operations o and ∇ on X and Y, respectively. Now the operations o and ∇ will be replaced by automata $\mathcal{C} = (X \times X,X,Z_C,z_{oC},\delta_C,\lambda_C)$ and $\mathcal{D} = (Y \times Y,Y,Z_D,z_{o,D},\delta_D,\lambda_D)$, respectively, with two input-terminals and one output-terminal each.
We call $\mathcal{C}$ a 2-input - 1-output automaton on X, shortly a (2-1)-automaton on X.
The set of all (2-1)-automata with input set $X \times X$ and output set X is denoted by $A_{2-1}(X)$.
Now a formal definition of $(\mathcal{C},\mathcal{D})$-superponability is given.

Definition 1. Let $\mathcal{A} = (X,Y,Z,z_o,\delta,\lambda)$, $\mathcal{C} \in A_{2-1}(X)$, $\mathcal{D} \in A_{2-1}(Y)$. Then $\mathcal{A}$ is called $(\mathcal{C},\mathcal{D})$-superponable if for all $t > 0$, $x_1,\ldots,x_t$, $x_1',\ldots,x_t' \in X$

$$g_t^{\mathcal{A}}(g_1^{\mathcal{C}}(x_1x_1'),\ g_2^{\mathcal{C}}(x_1x',x_2x_2'),\ldots,g_t^{\mathcal{C}}(x_1x_1',\ldots,x_tx_t')) = \qquad (1)$$
$$g_t^{\mathcal{D}}(g_1^{\mathcal{A}}(x_1)\ g_1^{\mathcal{A}}(x_1'),\ g_2^{\mathcal{A}}(x_1,x_2)\ g_2^{\mathcal{A}}(x_1',x_2'),\ldots,g_t^{\mathcal{A}}(x_1,\ldots,x_t)\ g_t^{\mathcal{A}}(x_1',\ldots,x_t')).$$

The automaton $\mathcal{A}$ is $(\mathcal{C},\mathcal{D})$-superponable if and only if the circuits represented in Fig. 1 are equivalent.

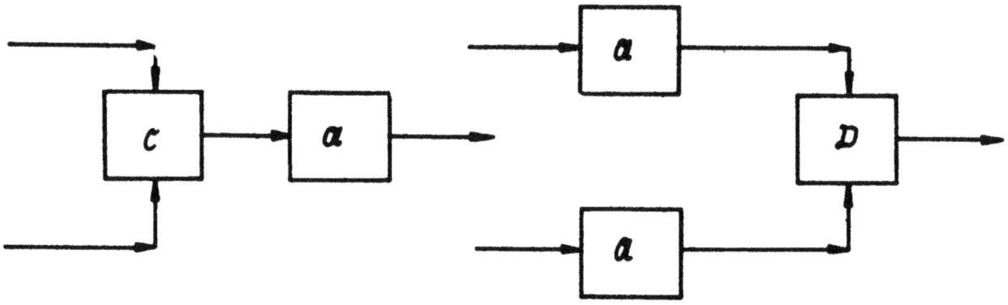

Fig.1

The circuits of Fig. 1 are called $\mathcal{C}-\mathcal{A}$-superposition and $(\mathcal{A},\mathcal{A})-\mathcal{D}$-superposition, respectively.

If $\mathcal{C}$ and $\mathcal{D}$ are combinational automata with

$$g_t^{\mathcal{C}}(x_1x_1',\dots,x_tx_t') = x_t \circ x_t' = F(x_t,x_t')$$

$$g_t^{\mathcal{D}}(y_1y_1',\dots,y_ty_t') = y_t \nabla y_t' = G(y_t,y_t')$$

for $t > 0$, $x_1,\dots,x_t,x_1',\dots,x_t' \in X$, $y_1,\dots,y_t,y_1',\dots,y_t' \in Y$, then the $(\mathcal{C},\mathcal{D})$-superponablity reduces to $(\circ,\nabla)$-superponability.

It is easy to convince oneself that for every automaton $\mathcal{A}$ there exist automata $\mathcal{C}_o$ and $\mathcal{D}_o$ such that $\mathcal{A}$ is $(\mathcal{C}_o,\mathcal{D}_o)$-superponable in a trivial sense.

Let $\mathcal{A} = (X,Y,Z,z_o,\delta,\lambda)$, $\mathcal{C}_o \in A_{2-1}(x)$ and $\mathcal{D}_o \in A_{2-1}(y)$.

For $t > 0$, $x_1,\dots,x_t,x_1',\dots,x_t' \in X$, $y_1,\dots,y_t,y_1',\dots,y_t' \in Y$ we define

$$g_t^{\mathcal{C}_o}(x_1x_1',\dots,x_tx_t') = \tilde{x}, \quad \tilde{x} \in X, \quad \tilde{x} \text{ fixed}$$

$$g_t^{\mathcal{D}_o}(y_1y_1',\dots,y_ty_t') = g_t^{\mathcal{A}}(\underbrace{\tilde{x},\dots,\tilde{x}}_{t}) \tag{2}$$

and $\mathcal{A}$ is obviously $(\mathcal{C}_o,\mathcal{D}_o)$-superponable but the output of the $\mathcal{C}-\mathcal{A}$ superposition depends on t only.

Furthermore every automaton $\mathcal{A}$ is even $(\circ,\nabla)$-superponable (i. e. with respect to operations) if $x \circ x' = x$ for $x,x' \in X$ and $y \nabla y' = y$ for $y,y' \in Y$.

Such superpositions are considered to be trivial.

In the following we are interested in nontrivial superpositions only.

A superposition is called nontrivial if its output at some t actually depends on x and x'.

<u>Definition 2.</u> Let $\mathcal{A}$ be $(\mathcal{C},\mathcal{D})$-superponable, then the $\mathcal{C}-\mathcal{A}$-superposition is called nontrivial if for some t, $t > 0$, there exist x_i and x_j', $1 \le i,j \le t$, such that $g_t^{\mathcal{C}-\mathcal{A}} : X^{2t} \to Y$ actually depends on x_i and x_j'.

The definition for the $(\mathcal{A},\mathcal{A})$-$\mathcal{D}$-superposition is quite similar. The following theorem deals with nontrivial superponability of automata. It is shown that every non-autonomous automaton is nontrivially superponable.

Theorem 1. Let $\mathcal{A} = (X,Y,Z,z_0,\delta,\lambda)$ be a non-autonomous automaton. Then there exist automata $\mathcal{C}$ and $\mathcal{D}$, $\mathcal{C} \in A_{2-1}(X)$ and $\mathcal{D} \in A_{2-1}(Y)$ such that

a) $\mathcal{A}$ is $(\mathcal{C},\mathcal{D})$-superponable

b) the $(\mathcal{A},\mathcal{A})$-$\mathcal{D}$-superposition (and therefore the $\mathcal{C}$-$\mathcal{A}$-superposition, too) is nontrivial.

Proof. The proof is by construction of $\mathcal{C}$ and $\mathcal{D}$. Let $Z_r = \{ z / \exists\, p \in X^*; z = \delta(z_0,p)\}$.
Since $\mathcal{A}$ is non-autonomous, there has to exist at least one state $\tilde{z},\tilde{z} \in Z_r$ such that

$$\lambda(z,p^1) \neq \lambda(z,p^2) \tag{3}$$

for some $p^1,p^2 \in X^*$ with $|p^1| = |p^2|$.
First we suppose that there exists at least one state z, $z \in Z_r$, such that

$$y = \lambda(\tilde{z},x) \neq \lambda(\tilde{z},x') = y' \tag{4}$$

for some $x,x' \in X$. (The case that there is no such state will be treated later.)
Let $Y_z = \{y / \exists\, x, \lambda(z,x) = y\}$. We attach to every $z \in Z_r$ and every $y \in Y_z$ exactly one $x_{z,y}$, $x_{z,y} \in X$ with $\lambda(z,x_{zy}) = y$ and define a function $\phi : Z \times Y \rightarrow X$ by

$$\phi(z,y) = x_{z,y} \qquad z \in Z_r,\quad y \in Y_z \tag{5}$$

with

$$\lambda(z(\phi(z,y))) = y \quad \text{if} \quad z \in Z_r,\quad y \in Y_z.$$

We denote by $\mathcal{A}'$ an automaton which describes the state transition of $\mathcal{A}$, $\mathcal{A}' = (X,Z,z_0,\delta)$.
Further we define a function $f : Y \times Y \times Z \rightarrow Y$ with the condition

$$f(y,y',z) = y'' \in Y_z. \tag{6}$$

By use of $\mathcal{A}$, $\mathcal{A}'$, ϕ and f we construct the automata $\mathcal{C}$ and $\mathcal{D}$.
In Fig. 2 the automaton $\mathcal{D}$ and the $(\mathcal{A},\mathcal{A})$-$\mathcal{D}$-superposition are represented.

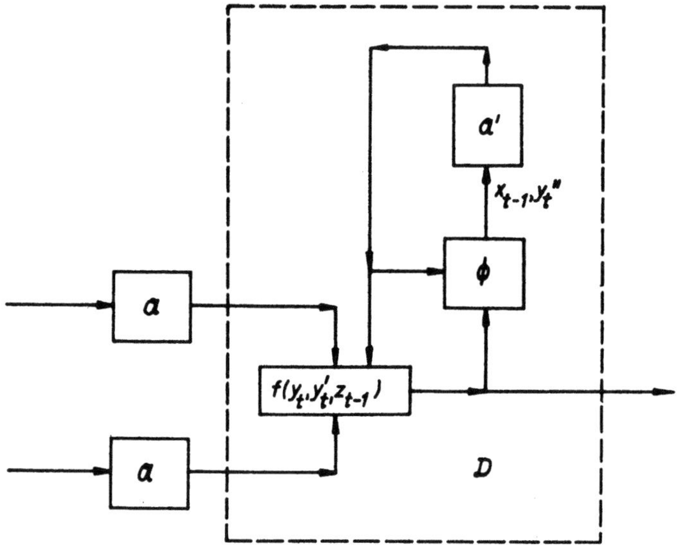

Fig.2

The automaton $\mathcal{C}$ and the $\mathcal{C}$-$\mathcal{a}$-superposition are shown in Fig. 3.

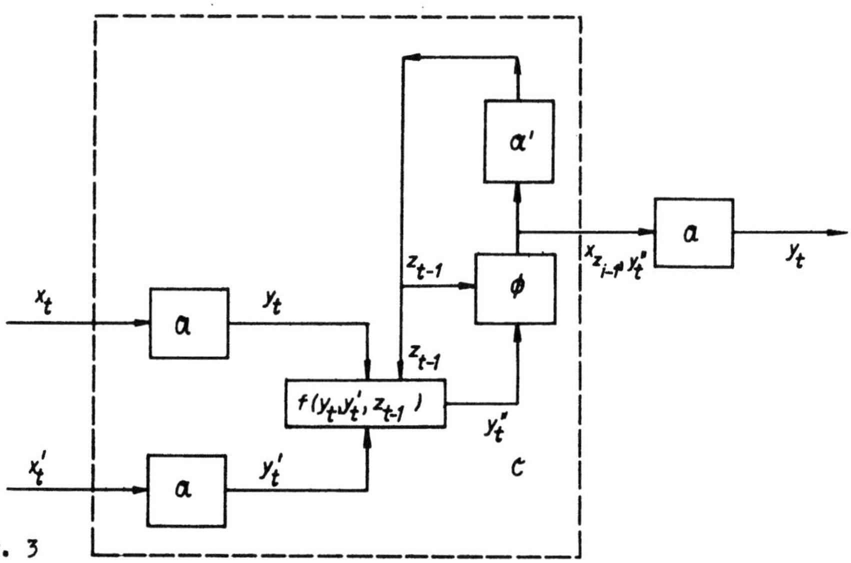

Fig. 3

Since $\lambda(z_{t-1}, x_{z_{t-1}}, y_t) = y_t$ and since $\mathcal{a}'$ and the automaton $\mathcal{a}$ on the right side of Fig. 3 are always in the same state, the $(\mathcal{a},\mathcal{a})$-$\mathcal{D}$-superposition and the $\mathcal{C}$-$\mathcal{a}$-superposition are equivalent.

Now we show that for a suitable choice of ϕ and f the $(\mathcal{a},\mathcal{a})$-$\mathcal{D}$-superposition actually depends on the inputs x and x'.

Since $\tilde{z} \in Z_r$, there has to exist at least one input word $x^1 \ldots x^k$ with $\tilde{z} = \delta(z_o, x^1 \ldots x^k)$. We suppose k to be minimal. Denoting $z^1 = \delta(z_o, x^1)$, $z^2 = \delta(z_o, x^1x^2), \ldots, z^{k-1} = \delta(z_o, x^1 \ldots x^{k-1})$, $\lambda(z_o, x^1 \ldots x^k) = y^1 \ldots y^k$ we have $z^i \not\sim z^j$ for $i \neq j$ and $z^i \not\sim \tilde{z}$ for $i,j = 0,1,\ldots,k-1$. Therefore it is possible to define without contradiction

$$\begin{aligned} &\phi(z^{i-1}, y^i) = x^i & i = 1,\ldots,k \\ &f(y^i, y^i, z^{i-1}) = y^i & i = 1,\ldots,h \\ &f(y,y,\tilde{z}) = y & \\ &f(y,y',\tilde{z}) = f(y',y,\tilde{z}) = f(y',y',\tilde{z}) = y' & \end{aligned} \tag{6}$$

and the output-sequences of the $(\mathfrak{a},\mathfrak{a})$-$\mathcal{D}$-superposition in response to the input-sequences $x_1 \ldots x_{k+1}$ and $x'_1 \ldots x'_{k+1}$:

1. $x^1 \ldots x^k x$ and $x^1 \ldots x^k x$,
2. $x^1 \ldots x^k x'$ and $x^1 \ldots x^k x'$,
3. $x^1 \ldots x^k x$ and $x^1 \ldots x^k x'$ and
4. $x^1 \ldots x^k x'$ and $x^1 \ldots x^k x$

are

1. $y^1 \ldots y^k y$, 2. $y^1 \ldots y^k y'$,
3. $y^1 \ldots y^k y'$ and 4. $y^1 \ldots y^k y'$, respectively.

Thus the $(\mathfrak{a},\mathfrak{a})$-$\mathcal{D}$-superposition is nontrivial.

Let us suppose now that for all $z \in Z_r$ and for all $x,x' \in X$ we have $\lambda(z,x) = \lambda(z,x')$. According to the suppositions of Theorem 1 there is at least on state $\tilde{z}$, $\tilde{z} \in Z_r$ such that $\lambda(\tilde{z},p^1) \neq \lambda(\tilde{z},p^2)$ for some $p^1,p^2 \in X^*$ with $|p^1| = |p^2|$. We choose $\tilde{z}$ and $p^1 = x_1^1 \ldots x_n^1$, $p^2 = x_1^2 \ldots x_n^2$, such that n is minimal.

Then we have for every $z \in Z_r$, $x_1,\ldots,x_n$, $x'_1,\ldots,x'_n \in X$

$$\begin{aligned} \lambda(z, x_1 \ldots x_n) &= y_1 \ldots y_n = y_1^z \ldots y_{n-1}^z \, y_n(x_1) \\ \lambda(z, x'_1 \ldots x'_n) &= y'_1 \ldots y'_n = y_1^z \ldots y_{n-1}^z \, y_n(x'_1). \end{aligned} \tag{7}$$

If $y_k \neq y'_k$, $k < n$, then n is not minimal. Therefore every state $z \in Z_r$ is associated with an output-sequence $y_1^z y_2^z \ldots y_{n-1}^z$ of length n-1 and every state senquence $z_1 \ldots z_k$ is associated with an output-sequence $y_1^{z_1} \ldots y_1^{z_k} y_2^{z_k} \ldots y_{n-1}^{z_k}$ of length k + n - 2. Thereby z_{i+1} is supposed to be a successor of z_i.

Instead of the functions f and ϕ we define functions $f_n : Y \times Y \times Z \longrightarrow Y$ and $\phi_n : Z \times Y \longrightarrow X$ with the conditions

$$f(y_t, y'_t, z_{t-1}) = y''_{t+n} \in Y_{n,z_{t-1}} \tag{8}$$

where

$$Y_{n,z_{t-1}} = \{y \,/\, \exists\, x_1,\dots,x_n \in X,\ z_{t-1} \in Z_r,\ y = \lambda(\delta(z_{t-1},x_1,\dots$$
$$\dots,x_{n-1}),\ x_n)$$

and

$$\phi(z_{t-1},\ y''_{t+n}) = x''_t, \tag{9}$$

where

$$y''_{t+n} = y_{n-1}^{\delta(z_{t-1},x''_t)}.$$

The automaton $\mathcal{D}$ and the $(\mathcal{A},\mathcal{A})$-$\mathcal{D}$-superposition are represented in Fig.4.

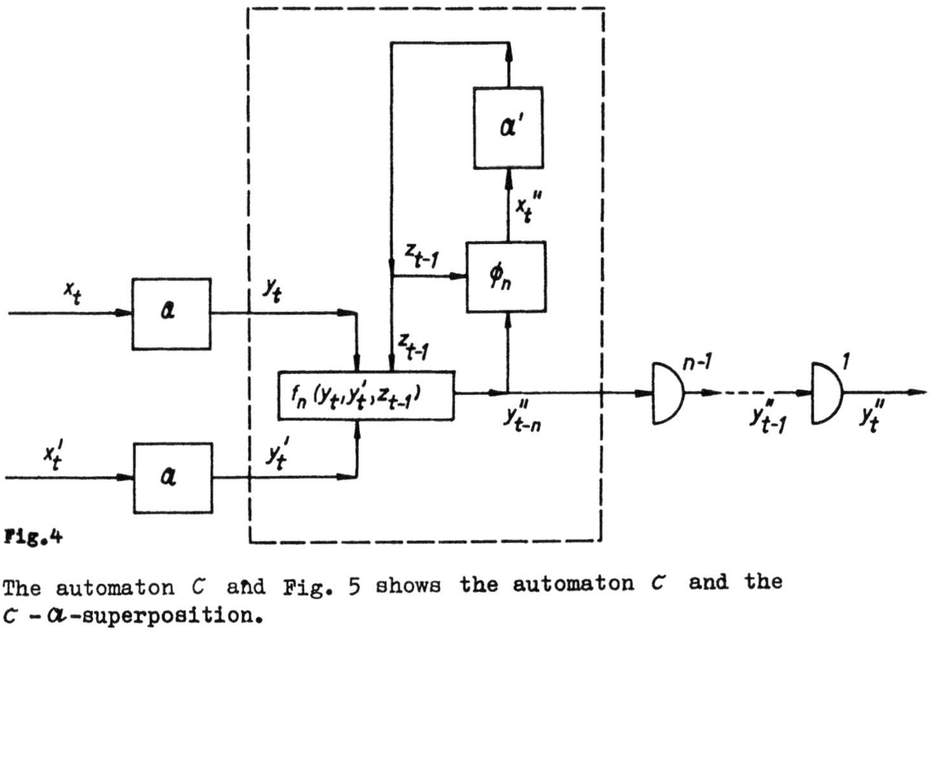

Fig.4

The automaton $\mathcal{C}$ and Fig. 5 shows the automaton $\mathcal{C}$ and the $\mathcal{C}$-$\mathcal{A}$-superposition.

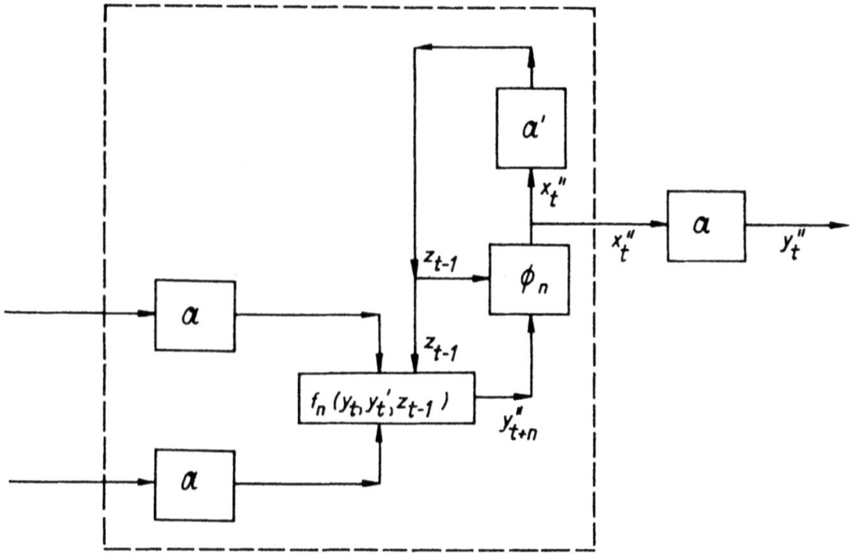

Fig.5

The state of $\mathfrak{a}$ at the right side of Fig. 5 and the state of $\mathfrak{a}'$ are identical. If $\mathfrak{a}$ is in state z_{t-1}, the next state is $z_t = \delta(z_{t-1}, x_t'')$ with $y''_{t+n} = y_{n-1}^{\delta(z_{t-1}, x_t'')}$. Therefore the output of the $(\mathfrak{a},\mathfrak{a})-\mathfrak{D}$-superposition and the $\mathcal{C}-\mathfrak{a}$-superposition for every t, t > 0, at time t + n coincide. If the initial states of the delay elements 1,...,n - 1 in Fig. 4 are $y_1^{z_o}, \ldots, y_{n-1}^{z_o}$, the $(\mathcal{C},\mathfrak{a})$-superposition and the $(\mathfrak{a},\mathfrak{a})-\mathfrak{D}$-superposition are equivalent.

Now we have to show that the superposition is nontrivial.
Let $\tilde{z} = \delta(z_o, \tilde{x}_1 \ldots \tilde{x}_k)$. We suppose k to be minimal.
Denoting $z_1 = \delta(z_o, \tilde{x}_1)$, $z_2 = \delta(z_o, \tilde{x}_1\, \tilde{x}_2), \ldots, z_{k-1} = \delta(z_o, \tilde{x}_1 \ldots \tilde{x}_{k-1})$
we have $z_i \not\sim z_j$ for $i \neq j$ and $z_i \not\sim \tilde{z}$ for $i,j = 0,\ldots,k-1$.
The output-sequences associated with the state sequences
$z_o \ldots z_{k-1}\, \tilde{z}\, \delta(z, x_1^1)$ and $z_o \ldots z_{k-1}\, \tilde{z}\, \delta(z, x_1^2)$ are
$\tilde{y}_1 \ldots \tilde{y}_k\, y_1^{\tilde{z}} \ldots y_{n-1}^{\tilde{z}}\, y^1$ and $\tilde{y}_1 \ldots \tilde{y}_k\, y_1^{\tilde{z}} \ldots y_{n-1}^{\tilde{z}}\, y^2$.
We define (without contradiction)

$$
\begin{aligned}
f_n(\tilde{y}_1, \tilde{y}_1, z_o) &= y_{n-1}^{z_1} \\
&\vdots \\
f_n(\tilde{y}_k, \tilde{y}_k, z_{k-1}) &= y_{n-1}^{\tilde{z}}
\end{aligned}
\qquad (10)
$$

$$f_n(y^1,y^1,\tilde{z}) = y^1$$
$$f_n(y^2,y^2,\tilde{z}) = f(y^2,y^1,\tilde{z}) = f(y^1,y^2,\tilde{z}) = y^2 \qquad (10)$$

$$\Phi_n(z_o,y^{z_1}_{n-1}) = x_1$$
$$\vdots$$
$$\Phi_n(z_{k-1}\ y^{\tilde{z}}_{n-1}) = x_k \qquad (11)$$
$$\Phi_n(\tilde{z},y^1) = x^1_1$$
$$\Phi_n(\tilde{z},y^2) = x^2_1 .$$

Then the input sequences $\tilde{x}_1 \ldots \tilde{x}_k\ x_{k+1} \ldots x_{k+n}$ and $\tilde{x}_1 \ldots \tilde{x}_k\ x'_{k+1} \ldots x'_{k+n}$ with

1. $x_{k+1} = x^1_1$ and $x'_{k+1} = x^1_1$, 2. $x_{k+1} = x^2_1$ and $x'_{k+1} = x^2_1$,

3. $x_{k+1} = x^1_1$ and $x'_{k+1} = x^2_1$, 4. $x_{k+1} = x^2_1$ and $x'_{k+1} = x^1_1$ imply

1. $y''_{k+n} = y^1$, 2. $y''_{k+n} = y^2$, 3. $y''_{k+n} = y^2$, 4. $y''_{k+n} = y^2$

and the superposition is nontrivial. ■

From the construction of $\mathcal{C}$ and $\mathcal{D}$ on the proof of Theorem 1 we imediately have

Conclusion 2. If $\mathfrak{A}$ is a finite non-autonomous automaton, then there exist finite automata $\mathcal{C}$ and $\mathcal{D}$, $\mathcal{C} \in A_{2-1}(X)$ and $\mathcal{D} \in A_{2-1}(Y)$, such that $\mathfrak{A}$ is nontrivially $(\mathcal{C},\mathcal{D})$-superponable.
Till now we have supposed $\mathcal{C} \in A_{2-1}(X)$ and $\mathcal{D} \in A_{2-1}(Y)$.

Now the automata $\mathcal{C}$ and $\mathcal{D}$ with two input-terminals and one output-terminal will be replaced by automata

$\mathcal{C}_n = (X^n,X,Z_i,z_{oi},\delta_C,\lambda_C)$ and $\mathcal{D}_n = (Y^n,Y,Z_D,z_{oD},\delta_D,\lambda_D)$, respectively.
$\mathcal{C}_n$ is called an n-input-1-output automaton on X, shortly an (n-1)-automaton on X.
The set of all (n-1)-automata on X is denoted by $A_{n-1}(X)$. A formal definition of $(\mathcal{C}_{n-1},\mathcal{D}_{n-1})$-superponability is given now.

Definition 3. Let $\mathfrak{A} = (X,Y,Z,z_o,\delta,\lambda)$, $\mathcal{C}_n \in A_{n-1}(X)$, $\mathcal{D}_n \in A_{n-1}(Y)$. Then $\mathfrak{A}$ is called $(\mathcal{C}_n,\mathcal{D}_n)$-superponable if for all $t > 0$,
$x^i_1,\ldots,x^i_t \in X$; $i = 1,\ldots,n$

$$g_t^{\mathfrak{A}}(g_1^{\mathcal{C}_n}(x_1^1\ldots x_1^n),\; g_2^{\mathcal{C}_n}(x_1^1\ldots x_1^n,\; x_2^1\ldots x_2^n),\ldots,g^{\mathcal{C}_n}(x_1^1\ldots x_1^n,\ldots,x_t^1\ldots x_t^n))$$
$$= g_t^{\mathcal{D}_n}(g_1^{\mathfrak{A}}(x_1^1)\ldots g_1^{\mathfrak{A}}(x_1^n),\; g_2^{\mathfrak{A}}(x_1^1,x_2^1)\ldots g_2^{\mathfrak{A}}(x_1^n,x_2^n),\ldots,g_t^{\mathfrak{A}}(x_1^1,\ldots,x_t^1)\ldots$$
$$\ldots g_t^{\mathfrak{A}}(x_1^n,\ldots,x_t^n)). \tag{12}$$

The automaton $\mathfrak{A}$ is $(\mathcal{C}_n,\mathcal{D}_n)$-superponable if the circuits represented in Fig. 6 are equivalent.

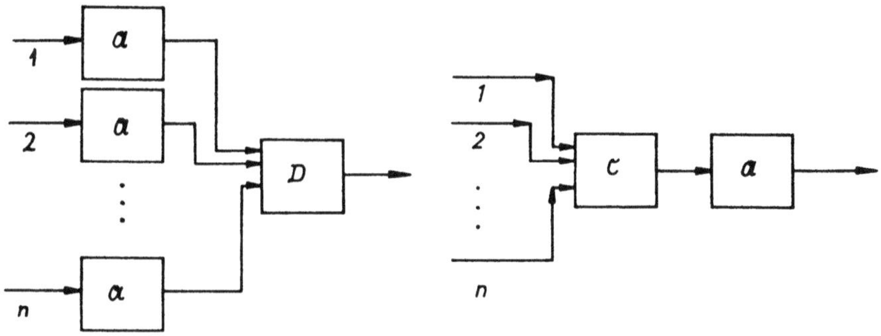

Fig.6

Similar to Theorem 1 we have

Theorem 3. Let $\mathfrak{A} = (X,Y,Z,z_0,\delta,\lambda)$ be a non-autonomous automaton. Then there exist automata $\mathcal{C}_n$, $\mathcal{D}_n$, $\mathcal{C}_n \in A_{n-1}(X)$ and $\mathcal{D}_n \in A_{n-1}(Y)$ such that

a) $\mathfrak{A}$ is $(\mathcal{C}_n,\mathcal{D}_n)$-superponable

b) the $(\mathfrak{A},\ldots,\mathfrak{A})$-$\mathcal{D}$-superposition is nontrivial.

The proof is omitted here.

Theorem 1 may be considered as one of the basic results concerning the superposition of (nonlinear) abstract automata. Since every automaton $\mathfrak{A}$ is $(\mathcal{C},\mathcal{D})$-superponable, it seems to be possible to classify nonlinear time-discrete systems modelled by abstract automata by their superposition properties. The problem is to find out such classes of automata with "interesting properties" for which theorems can be proved.
Till now linear automata, (o,∇)-linear automata, (F,G)-superponable binary automata and to some extend general (o,∇)-superponable automata have been characterized.
Now we inverstigate the superponability of invertible automata.

Definition 4. Let $\mathfrak{A} = (X,Y,Z,z_0,\delta,\lambda)$. Then $\mathfrak{A}$ is called invertible if there exists an automaton $\mathfrak{A}^{-1} = (Y,X,Z^{-1},z_0^{-1},\delta^{-1},\lambda^{-1})$ such that

1. for all $t > 0$, $x_1, \dots, x_t \in X$

$$g_t^{\mathcal{A}^{-1}}(g_1^{\mathcal{A}}(x_1),\ g_2^{\mathcal{A}}(x_1,x_2), \dots, g_t^{\mathcal{A}}(x_1,\dots,x_t)) = x_t \qquad (13)$$

2. for all $t > 0$, $y_1, \dots, y_t \in Y$

$$g_t^{\mathcal{A}}(g_t^{\mathcal{A}^{-1}}(y_1),\ g_t^{\mathcal{A}^{-1}}(y_1,y_2), \dots, g_t^{\mathcal{A}^{-1}}(y_1,\dots,y_t)) = y_t \qquad (14)$$

$\mathcal{A}^{-1}$ is called an inverse automaton of $\mathcal{A}$. $\mathcal{A}^{-1}$ is determined up to equivalence.

The definition 4 is symmetric with respect to $\mathcal{A}$ and $\mathcal{A}^{-1}$. The serial connection of $\mathcal{A}$ and $\mathcal{A}^{-1}$ is the identical mapping from X to X and the serial connection of $\mathcal{A}^{-1}$ and $\mathcal{A}$ is the identical mapping from Y to Y.

If the automaton $\mathcal{A}$ is invertible, then it is possible to determine all pairs of automata $\mathcal{C}$, $\mathcal{D}$ for which $\mathcal{A}$ is $(\mathcal{C},\mathcal{D})$-superponable.

Theorem 4. Let $\mathcal{A} = (X,Y,Z,z_0,\delta,\lambda)$ be an invertible automaton and let $\mathcal{A}^{-1}$ be an inverse automaton of $\mathcal{A}$.
Then $\mathcal{A}$ is $(\mathcal{C},\mathcal{D})$-superponable if and only if

1. $\mathcal{C} \in A_{2-1}(X)$ and $\mathcal{D} \sim (\mathcal{A}^{-1},\mathcal{A}^{-1})-\mathcal{C}-\mathcal{A}$

or

2. $\mathcal{D} \in A_{2-1}(Y)$ and $\mathcal{C} \sim (\mathcal{A},\mathcal{A})-\mathcal{D}-\mathcal{A}^{-1}$.

Proof. 1a) Let $\mathcal{A}$ be $(\mathcal{C},\mathcal{D})$-superponable. Then $\mathcal{C}-\mathcal{A}$ and $(\mathcal{A},\mathcal{A})-\mathcal{D}$ are equivalent. By a serial connection of both automata circuits with $\mathcal{A}^{-1}$ we conclude that $\mathcal{C}$ and $(\mathcal{A},\mathcal{A})-\mathcal{D}-\mathcal{A}^{-1}$ are equivalent.
b) Let $\mathcal{C} \sim (\mathcal{A},\mathcal{A})-\mathcal{D}-\mathcal{A}^{-1}$. By a serial connection of both automata circuits with $\mathcal{A}$ we conclude $\mathcal{C}-\mathcal{A} \sim (\mathcal{A},\mathcal{A})-\mathcal{D}$, i. e. $\mathcal{A}$ is $(\mathcal{C},\mathcal{D})$-superponable.
The second statement can be proved quite similarly. ■

Now it is possible to ask under which conditions the automata $\mathcal{C}$ and $\mathcal{D}$ or one of them can be chosen with a finite input memory or a finite input-output memory. The answer to this question is still quite incomplete. We discuss here only the problem under which conditions it is possible to choose the automaton $\mathcal{D}$, $\mathcal{D} \in A_{2-1}(Y)$ as a combinational one, i. e. as an operation. (If $\mathcal{C}$ and $\mathcal{D}$ are both combinational automata, s. chapter 3.)

First let us consider the automaton $\mathcal{A} = (X = \{x_1,x_2\},\ Y = \{y_1,y_2\},\ Z = \{z_1,\dots,z_8\},\ z_1,\ \delta,\ \lambda)$ with the diagram represented in Fig. 7.

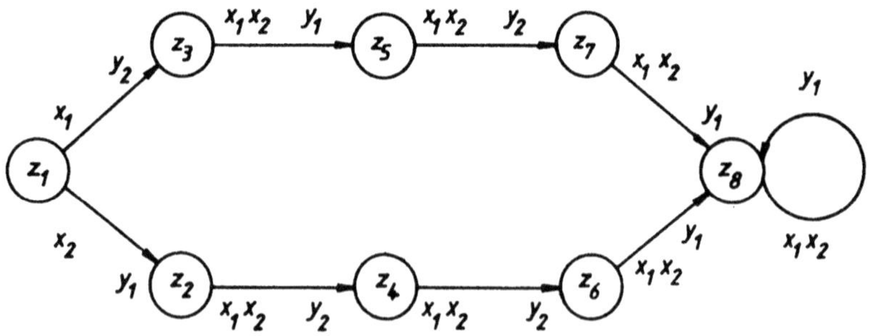

Fig.7

Let $\mathfrak{A}$ be $(\mathcal{C},\nabla)$-superponable where ∇ is a 2-ary operation on Y and $\mathcal{C} \in A_{2-1}(X)$. Then we necessarily have either $y \nabla y' = y$ or $y \nabla y' = y'$ for $y,y' \in Y$, i. e. the $(\mathcal{C},\nabla)$-superposition is trivial. ($y_2 \nabla y_1 = y_1$ implies $y_1 \nabla y_2 = y_2$, $y_2 \nabla y_2 = y_2$, $y_1 \nabla y_1 = y_1$, i. e. $y \nabla y' = y'$ for $y,y' \in Y$; and $y_2 \nabla y_1 = y_2$ implies $y_1 \nabla y_2 = y_1$, $y_2 \nabla y_2 = y_2$, $y_1 \nabla y_1 = y_1$, i. e. $y \nabla y' = y$ for $y,y' \in Y$.)

This example shows that there exist automata not $(\mathcal{C},\nabla)$-superponable, where ∇ is a 2-ary operation. A sufficient condition for the existence of an operation ∇ such that an automaton $\mathfrak{A}$ is $(\mathcal{C},\nabla)$-superponable is formulated in Theorem 5.

<u>Theorem 5.</u> Let $\mathfrak{A} = (X,Y,Z,z_o,\delta,\lambda)$ and let $Y_{Z_r} = \bigcap_{z \in Z_r} Y_z$ be the set of outputs possible in every state of Z_r. If $|Y_{Z_r}| > 1$, then there exist a 2-ary operation ∇ on Y and an automaton $\mathcal{C} \in A_{2-1}(X)$ such that $\mathfrak{A}$ is $(\mathcal{C},\nabla)$-superponable and the superposition is nontrivial.

Proof. Since $|Y_{Z_r}| > 1$, we may define in the proof of Theorem 1

$$f(y,y',z) = y \nabla y' \in Y_{Z_r}$$

independent of z and the automaton $\mathfrak{A}'$ in Fig. 2 may be omitted. Then the automaton $\mathfrak{D}$ is a combinational one. ■

If $\mathfrak{A}$ is invertible, then we have $Y_z = Y$ for all $z \in Z_r$ and therefore $Y_{Z_r} = Y \neq 0$. From Theorem 4 and Theorem 5 we conclude

<u>Conclusion 6.</u> Let $\mathfrak{A} = (X,Y,Z,z_o,\delta,\lambda)$ be an invertible automaton and let ∇ be an arbitrary 2-ary operation on Y. Then $\mathfrak{A}$ is $(\mathcal{C},\nabla)$-superponable with $\mathcal{C} = (\mathfrak{A},\mathfrak{A}) - \nabla) - \mathfrak{A}^{-1}$.

A similar statement can be proved for an arbitrary 2-ary operation o on

X.

The following example shows that the conditions of Theorem 5 are not necessary.

Example. Let $\mathfrak{A} = (X = \{x_1, x_2\},\ Y = \{y_1, y_2, y_3\},\ Z = \{z_1, z_2, z_3\},\ z_1, \delta, \lambda)$ with the state diagram of Fig. 8.

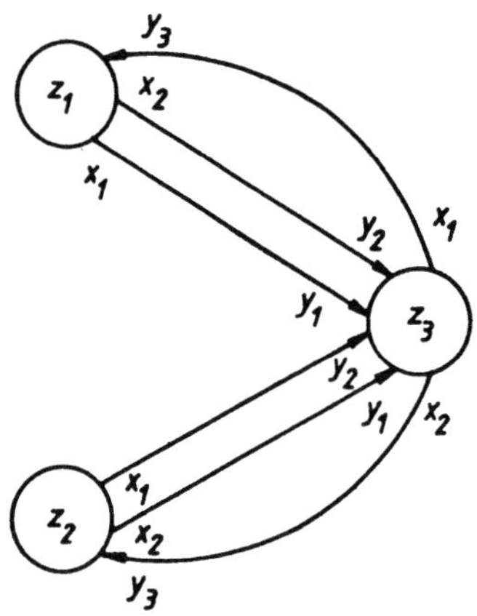

Fig.8

We have $Y_{Z_r} = \emptyset$ but it is possible to define ∇ on Y by

$y_1 \nabla y_1 = y_2 \nabla y_2 = y_1$, $y_1 \nabla y_2 = y_2 \nabla y_1 = y_2$, $y_3 \nabla y_3 = y_3$ and arbitrarily in all the other cases. If t is odd, then $\mathfrak{A}$ is in state z_3, if t is even, then $\mathfrak{A}$ may be either in state z_1 or in state z_2. Some other more detailed theorems can be proved but till now there have not been known necessary and sufficient conditons for the existence of a 2-ary operation ∇ on Y such that a given automaton is $(\mathcal{C}, \nabla)$-superponable for some $\mathcal{C}$.

At the end of this chapter we discuss two examples of $(\mathcal{C}, \mathcal{D})$-superponable automata.

As a first simple example let us consider a binary modulo 2 adder $\mathfrak{A}_1 = (X, Y, Z, z_o, \delta, \lambda)$ with $X = Y = Z = \{0,1\}$, $z_o = 0$, $\delta(z,x) = z + x$ (modulo 2), $\lambda(z,x) = z + x$ (modulo 2). The input-output-behaviour of $\mathfrak{A}$ is described by

$$y_t = \sum_{i=1}^{t} x_i \qquad \text{(modulo 2)}, \tag{15}$$

$\mathfrak{A}_1$ is invertible with $\mathfrak{A}_1^{-1} = (Y, X, Z, z_o, \delta^{-1}, \lambda^{-1})$, $\delta^{-1}(z,y) = y$, $\lambda^{-1}(z,y) = z + y$ (modulo 2) and the input-output-behaviour of $\mathfrak{A}^{-1}$ is

$$x_t = y_t + y_{t-1} \quad \text{(modulo 2)}. \tag{16}$$

A realization of $\mathfrak{A}_1$ and $\mathfrak{A}_1^{-1}$ is shown in Fig. 9.

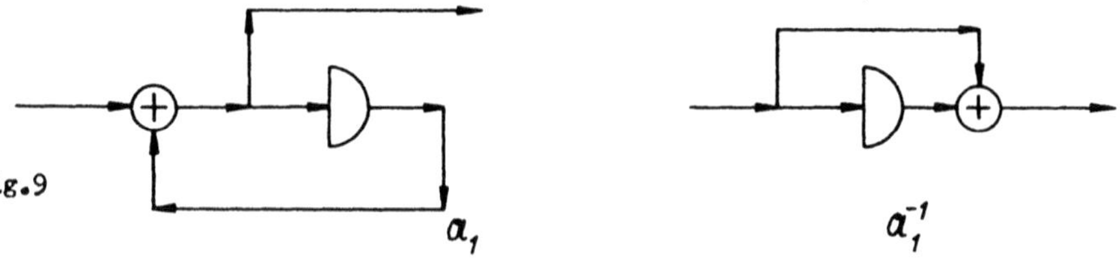

Fig.9

Since $\mathfrak{A}_1$ is linear (over GF(2)), it is (+,+)-superponable. Since $\mathfrak{A}_1$ is invertible, $\mathfrak{A}_1$ is superponable with respect to an arbitrary $\mathcal{D} \in A_{2-1}(\{0,1\})$ and $\mathcal{C} \sim (\mathfrak{A}_1, \mathfrak{A}_1) - \mathcal{D} - \mathfrak{A}_1^{-1}$ (Theorem 4). By Conclusion 6 $\mathcal{D}$ can be chosen as an arbitrary 2-ary Boolean function G(y,y').

Let G(y,y') = y $\vee$ y' (disjunction). The corresponding automaton $\mathcal{C}^{\vee}$ is represented in Fig. 10.

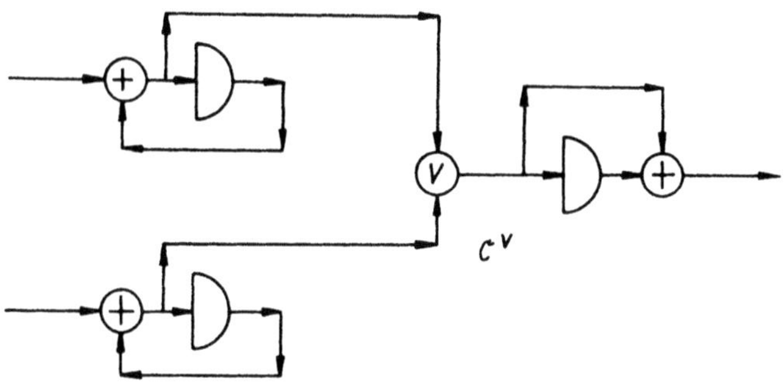

Fig.10

$\mathfrak{A}_1$ is obviously $(\mathcal{C}^{\vee}, \vee)$-superponable.
As a second example let us consider the automaton $\mathfrak{A}_2 = (X,Y,Z,z_o,\delta,\lambda)$ with X = Y = $\{0,1\}$, Z = $\{[0,0], [0,1], [1,0], [1,1]\}$, z_o = [0,0] with finite input-output-memory of the lenght 1 and the response formula (+ modulo 2)

$$y_t = x_t \vee x_{t-1} + y_{t-1}, \quad x_o = y_o = 0. \tag{17}$$

Fig. 11 shows a realization and the state diagram of $\mathfrak{A}_2$.

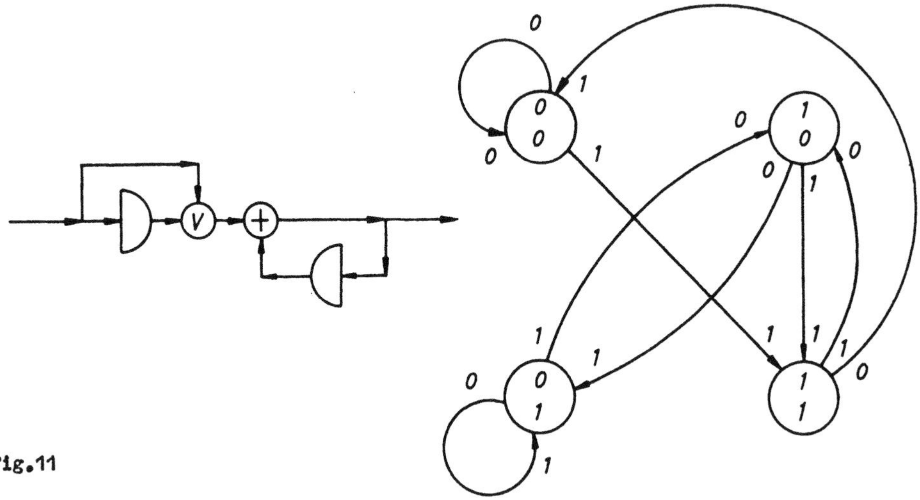

Fig.11

$\mathfrak{A}$ is not invertible. But there is a state z, e. g. $\tilde{z} = [0,1]$ such that $1 = \lambda([0,1], 0) \neq \lambda([0,1], 1) = 0$, and the automata $\mathfrak{D}$ and $\mathcal{C}$ can be constructed as illustrated in Fig. 2 and 3, respectively.
We have $Y_{[0,0]} = Y_{[0,1]} = \{0,1\}$, $Y_{[1,0]} = \{1\}$, $Y_{[1,1]} = \{0\}$. The function $\phi : Z \times Y \rightarrow X$ can be defined by

$$\phi([0,0], 0) = 0, \quad \phi([0,0], 1) = 1, \quad \phi([0,1], 0) = 1,$$
$$\phi([0,1], 1) = 0, \quad \phi([1,0], 1) = 0, \quad \phi([1,1], 0) = 1.$$

The function $f : Y \times Y \times Z \rightarrow Y$ can be defined by

$$f(y,y', [0,0]) = f(y,y', [0,1]) = y + y', \quad f(y,y', [1,0]) = 1,$$
$$f(y,y', [1,1]) = 0 \quad \text{for } y,y' \in Y$$

and the automaton $\mathfrak{A}' = (X,Z,z_0,\delta)$ describes the state behaviour of $\mathfrak{A}$.

5. Invariant relations of automata

In this chapter we discuss the conception of invariant relations for automata, which is introduced in /19/.
Automata are open systems with respect to their inputs and outputs. Therefore the invariants of automata have to reflect and to describe properties of their input-output-behaviour.
The conception of invariant relations of automata is based on the theory of functions preserving a relation on a finite set (for details s. /6, 35, 38/).
Here the functions under consideration are the input-output functions $g_t^{\mathfrak{A}}$ of an automaton $\mathfrak{A}$.
Since the conception of invariant relations includes as a special case the generalized principle of the superposition of automata with respect to operations, it is to be considered as a natural extension of this principle. Many other well-known domains of automata theory like homomorphisms of automata, linear automata and tolerance automata are also within this conception.
It is of special interest to notice here that it is possible to extend the description of the input-output-behaviour by means of generalized impulse responses to a large class of nonlinear automata preserving a pair of relations.
Problems concerning the approximation of an automaton by other ones can be treated.
We have already mentioned that we present here a conception and not a theory of invariant relations of automata.
This chapter is essentially based on /19/, where stochastic and non-deterministic automata, properties of relations invariant for some automata and some other problems are treated.

First we introduce some notions and definitions.
Let X be a set and let $R_X^{(m)}$ denote the set of all m-ary relations on X where $m > 0$. Then each relation $\rho \in R_X^{(m)}$ is a subset of X^m. The elements $r \in \rho$ of an m-ary relation are described as m-dimensional columnvectors r, where r(i), i = 1,...,m, is the i-th component of r,

$$r = \begin{pmatrix} r(1) \\ \vdots \\ r(m) \end{pmatrix} = [r(1),\dots,r(m)].$$

The invariance of a pair of relations with respect to a function is explained by the following definition.

Definition 1. A pair of relations (ϱ,σ), $\varrho \in R_X^{(m)}$, $\sigma \in R_Y^{(m)}$ is called to be invariant for an n-ary function $f : X^n \to Y$, or f is said to preserve (ϱ,σ) if

$$f(r_1,\ldots,r_n) = [f(r_1(1),\ldots,r_m(1)),\ldots,f(r_n(1),\ldots,r_n(m))] \qquad (1)$$

belongs to σ if $r_1,\ldots,r_n \in \varrho$.

Since every automaton $\mathfrak{A}$ is associated with a set $g_t^{\mathfrak{A}}$, $t = 1,2,3,\ldots$ of t-ary functions, it is possible to extend Definition 1 to automata or more general to m-tuples of automata.

Definition 2. Let $(\mathfrak{A}_1,\ldots,\mathfrak{A}_m)$ be an m-tuple of automata.

$\mathfrak{A}_i = (X,Y,Z_i,z_{oi},\delta_i,\lambda_i)$ $i = 1,\ldots,m$ and $m > 0$,

$\varrho \in R_X^{(m)}$, $\sigma \in R_Y^m$.

Then we say $(\mathfrak{A}_1,\ldots,\mathfrak{A}_m)$ T-preserves (ϱ,σ), or (ϱ,σ) is T-invariant for $(\mathfrak{A}_1,\ldots,\mathfrak{A}_m)$ if for all t, $0 < t \leq T$ and for all $r_1,\ldots,r_t \in \varrho$

$$[g_t^{\mathfrak{A}_1}(r_1(1),\ldots,r_t(1)),\ldots,g_t^{\mathfrak{A}_m}(r_1(m),\ldots,r_t(m))] \in \sigma . \qquad (2)$$

If $(\mathfrak{A}_1,\ldots,\mathfrak{A}_m)$ T-preserves (ϱ,σ) for all $T > 0$, then $(\mathfrak{A}_1,\ldots,\mathfrak{A}_m)$ is said to preserve (ϱ,σ) (or (ϱ,σ) is invariant).

If $\mathfrak{A}_1 = \mathfrak{A}_2 = \ldots = \mathfrak{A}_m = \mathfrak{A}$, then we say that $\mathfrak{A}$ T-preserves (or preserves) (ϱ,σ).

The definition of (ϱ,σ)-invariance of $(\mathfrak{A}_1,\ldots,\mathfrak{A}_m)$ is illustrated in Fig. 1. Elements of ϱ are transformed by $(\mathfrak{A}_1,\ldots,\mathfrak{A}_m)$ to elements of σ if $(\mathfrak{A}_1,\ldots,\mathfrak{A}_m)$ preserves (ϱ,σ).

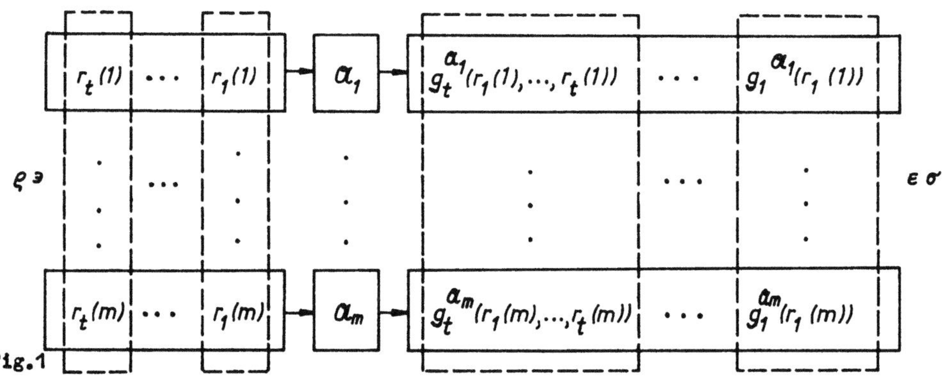

Fig.1

It is easy to see that some well-known areas of automata theory are contained in the conception of invariant relations. We show this for linear automata, generalized principles of superposition, homomorphisms of automata, and tolerance automata.

1. Linear automata. Let $\mathfrak{A} = (X,Y,Z,z_0,\delta,\lambda)$ with $X = Y = K$, where K is a given field $(K,+,.)$.
Defining $\varrho = \sigma = \{[x,x',x + x'] / x,x' \in X\}$ we have:
$\mathfrak{A}$ is $(+,+)$-superponable, i. e. linear (cf. chapter 3, Theorem 14) if (ϱ,σ) is invariant for $\mathfrak{A}$.

2. Superposition principles for automata. Let $(\mathfrak{A}_1,\dots,\mathfrak{A}_m,\mathfrak{B})$ be an $(m + 1)$-tuple of automata, $\mathfrak{A}_i = (X,Y,Z_i,z_{oi},\delta_i,\lambda_i)$, $(i = 1,\dots,m)$; $\mathfrak{B} = (X,Y,Z,z_o,\delta,\lambda)$, let $\alpha : X^m \to X$, $\beta : Y^m \to Y$ be m-ary functions.
Defining

$$\begin{aligned} \alpha^o &= \{[x_1,\dots,x_m, \alpha(x_1,\dots,x_m)]\} \\ \beta^o &= \{[y_1,\dots,y_m, \beta(y_1,\dots,y_m)]\}, \end{aligned} \tag{3}$$

then (α^o,β) is invariant for $(\mathfrak{A}_1,\dots,\mathfrak{A}_m,\mathfrak{B})$ if and only if for all $t > 0$, $x_{11},\dots,x_{1t},\dots,x_{m1},\dots,x_{mt} \in X$

$$\begin{aligned} &\beta(g_t^{\mathfrak{A}_1}(x_{11},\dots,x_{1t}),\dots,g_t^{\mathfrak{A}_m}(x_{m1},\dots,x_{mt})) = \\ &= g_t^{\mathfrak{B}}(\alpha(x_{11},\dots,x_{m1}),\dots, \alpha(x_{1t},\dots,x_{mt})) \end{aligned} \tag{4}$$

We say $(\mathfrak{A}_1,\dots,\mathfrak{A}_m,\mathfrak{B})$ is (α,β)-superponable.

For $m = 2$, $\mathfrak{A}_1 = \mathfrak{A}_2 = \mathfrak{B}$ we obtain the (α,β)-superponability of $\mathfrak{A}$ (chapter 3.1., Definition 1).

For $m = 1$ we get:

Let $\mathfrak{A} = (X,Y,Z',z_o',\delta',\lambda')$ and $\mathfrak{B} = (X,Y,Z,z_o,\delta,\lambda)$, $\alpha : X \to X$, $\beta : Y \to Y$ unary functions.

Then $(\mathfrak{A},\mathfrak{B})$ preserves (α^o,β^o) if (α,β,γ) with $\gamma : Z' \to Z$ is an (α,β,γ) homomorphism from $\mathfrak{A}$ to $\mathfrak{B}$.

3. Tolerance automata. Let $\varrho \in R_X^{(2)}$, $\sigma \in R_Y^{(2)}$ be tolerance relations, i. e. reflexive and symmetric relations.

Let $\varrho^* \in R_X^2$, $\sigma^* \in R_Y^2$ be tolerance relations induced by ϱ and σ:

$[p,p'] \in \varrho^*$ for $p = (x_1 \dots x_n)$, $p' = (x_1' \dots x_m')$ if and only if $n = m$ and $[x_i,x_i'] \in \varrho$, $i = 1,\dots,n$ (analogous for σ^*).

Then we have:

$\mathfrak{A}$ is an input-output-tolerance automaton with the tolerance spaces

(X^*, ϱ^*) and (Y^*, σ^*) if and only if $\mathfrak{A}$ preserves (ϱ, σ). (For details s. /3/.)

If the automata preserving a pair of relations (ϱ, σ) are finite, the invariance of (ϱ, σ) can be reduced to T-invariance for some T.

In this case the process of checking the invariance of (ϱ, σ) becomes finite.

<u>Theorem 1.</u> Let $(\mathfrak{A}_1, \ldots, \mathfrak{A}_m)$ with $\mathfrak{A}_i = (X, Y, Z_i, z_{oi}, \delta_i, \lambda_i)$, $|Z_i| = n_i$, $n_i > 0$, be an m-tuple of finite automata and let $T = n_1 n_2 \ldots n_m$, $\varrho \in R_X^m$, $\sigma \in R_Y^m$. Then $(\mathfrak{A}_1, \ldots, \mathfrak{A}_m)$ preserves (ϱ, σ) if and only if $(\mathfrak{A}_1, \ldots, \mathfrak{A}_m)$ T-preserves (ϱ, σ).

Proof. We show: $(\mathfrak{A}_1, \ldots, \mathfrak{A}_m)$ T-preserves (ϱ, σ) implies $(\mathfrak{A}_1, \ldots, \mathfrak{A}_m)$ preserves (ϱ, σ). The other direction is trivial. We define an automaton $\mathfrak{A} = (\underline{X}, \underline{Y}, \underline{Z}, \underline{z}_o, \delta_{\mathfrak{A}}, \lambda_{\mathfrak{A}})$ with

$$\underline{X} = \{\underline{x} / \underline{x} \in \varrho\} \subseteq X^m,\ Y \subseteq Y^m,\ \underline{Z} = Z_1 x \ldots x Z_m,\ \underline{z}_o = [z_{o1}, \ldots, z_{om}],$$

$$\delta_{\mathfrak{A}}(\underline{z}, \underline{x}) = [\delta_1(z_1, x_1), \ldots, \delta_m(z_m, x_m)]$$

$$\lambda_{\mathfrak{A}}(\underline{z}, \underline{x}) = [\lambda_1(z_1, x_1), \ldots, \lambda_m(z_m, x_m)],$$

where $\underline{x} = [x_1, \ldots, x_m] \in \underline{X}$, $\underline{z} = [z_1, \ldots, z_m]$.

$\mathfrak{A}$ is the parallel composition of $\mathfrak{A}_1, \ldots, \mathfrak{A}_m$ where the input set X^m is restricted to $\underline{X}$.

It is evident by the construction of $\mathfrak{A}$ that $(\mathfrak{A}_1, \ldots, \mathfrak{A}_m)$ T-preserves (ϱ, σ) if and only if for all t, $0 < t \leq T$ and for all $\underline{p} \in \underline{X}^t$

$$g^{\mathfrak{A}}_{|\underline{p}|}(\underline{p}) \in \sigma \quad .$$

Thus we only have to show that $g^{\mathfrak{A}}_{|p|}(\underline{p}) \in \sigma$ for all $\underline{p} \in \underline{X}^{|\underline{p}|}$ with $|p| \leq T$ implies $g^{\mathfrak{A}}_{|\underline{p}|}(\underline{p}) \in \sigma$ for all $\underline{p} \in \underline{X}^*$.

The number of non-equivalent states of $\mathfrak{A}$ is at most $n_1 n_2 \ldots n_m = T$. Let $\underline{z} = \delta(\underline{z}_o, p^1)$ for $\underline{p}^1 \in \underline{X}^*$ with $|\underline{p}^1| > T$. Then there exists at least one input word $\underline{p}^2 \in \underline{X}^*$ with $|\underline{p}^2| < T$ such that $\underline{z} = \delta_{\mathfrak{A}}(\underline{z}_o, \underline{p}^2)$, which implies for $\underline{x} \in \varrho$

$$g^{\mathfrak{A}}_{|\underline{p}^1|+1}(\underline{p}^1 \underline{x}) = g^{\mathfrak{A}}_{|\underline{p}^2|+1}(\underline{p}^2 \underline{x}) \in \sigma \quad ,$$

which finishes the proof.

From Theorem 1 we obtain immediately

<u>Conclusion 2.</u> Let $\mathfrak{A} = (X, Y, Z, z_o, \delta, \lambda)$ with $|Z| = n$ be a finite automaton, o and $\triangledown$ 2-ary operations on X and Y, respectively. Then $\mathfrak{A}$ is $(o, \triangledown)$-superponable if and only if for all t, $0 < t \leq n^2$,

$x_1,\dots,x_t,x'_t,\dots,x'_t \in X$ we have

$$g_t^{\alpha}(x_1 \circ x'_1,\dots,x_t \circ x'_t) = g_t^{\alpha}(x_1,\dots,x_t) \triangledown g_t^{\omega}(x'_1,\dots,x'_t). \qquad (5)$$

The following definitions and results deal with the description of the input-output-behaviour of automata preserving pairs of relations by means of generalized impulse responses "up to relational exactness".

Definition 3. Let $\varrho \in R_X^{(m)}$ and let $\pi : \{1,\dots,m\} \to \{1,\dots,m\}$ be a permutation. Then the relation $\pi\varrho \in R_X^{(m)}$ is defined by

$$[x_1,\dots,x_m] \in \pi\varrho \quad \text{iff} \quad [x_{\pi(1)},\dots,x_{\pi(m)}] \in \varrho.$$

Definition 4. Let $\varrho \in R_X^{(m)}$ and $n > 0$. A set $B \subseteq X^n$ is called a ϱ-generator of X^n if for $\underline{x}^o \in X^n$ either $\underline{x}^o \in B$ or, if there exist $m-1$ elements, $\underline{x}^1,\dots,\underline{x}^{m-1} \in B$ such that

$$[\underline{x}^o(1),\underline{x}^1(1),\dots,\underline{x}^{m-1}(1)],\dots,[\underline{x}^o(n),\underline{x}^1(n),\dots,\underline{x}^{m-1}(n)] \in \pi\varrho$$

for some permutation π of $\{0,\dots,m-1\}$.

Definition 5. A ϱ-generator of X^n is called a ϱ-basis of X^n if no subset of B is a ϱ-generator of X^n. A ϱ-basis is called minimal if there is no ϱ-basis of X^n with fewer than $|\varrho|$ elements.

Sometimes it is possible to derive a ϱ-generator of X^n by use of a ϱ-generator of X^k, $k < n$. An example of such a case is a totally symmetric relation.
(If $\varrho \in R_X^{(m)}$ is a totally symmetric relation, then $[x_1,\dots,x_m] \in \varrho$ implies $[x_{\pi(1)},\dots,x_{\pi(m)}] \in \varrho$ for all permutations π.)
Then we have:

Theorem 3. Let $\varrho \in R_X^{(m)}$ be a totally symmetric relation, $B \subseteq X$ a ϱ-generator of X and $[b,\dots,b] \in \varrho$ for all $b \in B$.
Then B^n is a ϱ generator of X^n.

Proof. Let $\underline{x}^o = (x^o(1) \dots x^o(n)) \in X^n$. Since B is a ϱ-generator of X for every $x^o(j)$, $j = 1,\dots,n$, there exist $x^i(j) \in B$, $i = 1,\dots,m-j$, such that $[x^o(j),x^1(j),\dots,x^{m-1}(j)] \in \varrho$.
(For $x^o(j) = b \in B$ put $x^i(j) = b$) and therefore we have

$$[x^o(1),x^1(1),\dots,x^{m-1}(1)],\dots,[x^o(n),x^1(n),\dots,x^{m-1}(m)] \in \varrho. \blacksquare$$

The next theorem shows us how to describe the input-output-behaviour of automata preserving a pair of relations.

Theorem 4. Let $\mathfrak{A} = (X,Y,Z,z_0,\delta,\lambda)$ be an automaton preserving (ϱ,σ), $\varrho \in R_X^{(m)}$, $\sigma \in R_Y^{(m)}$ and let $B \subseteq X^n$ be a ϱ-generator of X^n. Then the output words $\lambda(b) = y^b(1) \ldots y^b(n)$, $b \in B$, completely describe the input-output-behaviour of $\mathfrak{A}$ "up to (ϱ,σ)", i. e. for all output words $\lambda(x_1 \ldots x_n) = y(1) \ldots y(n)$, $(x_1,\ldots,x_n \in X)$ there exist $b_1,\ldots,b_{m-1} \in B$ such that

$$[y(1),y^{b_1}(1),\ldots,y^{b_{m-1}}(1)],\ldots,[y(n),y^{b_1}(n),\ldots,y^{b_{m-1}}(n)] \in \pi\sigma$$

for some permutation π.
$\lambda(b)$, $b \in B$ is called the set of generalized impulse responses.

Proof. The proof is obvious, because $\mathfrak{A}$ preserves (ϱ,σ) and by the definition of a ϱ-generator. ■

Let us illustrate Theorem 4 by a simple example.

Example. Let $\mathfrak{A} = (X,Y,Z,z_0,\delta,\lambda)$ be an automaton with $X = Y = Z = \{0,1,2\}$, $\delta(z,x) = \lambda(z,x) = \max(z,x)$, $z_0 = 0$, $\varrho = \sigma = \left(\begin{pmatrix}1\\0\end{pmatrix},\begin{pmatrix}2\\0\end{pmatrix},\begin{pmatrix}2\\1\end{pmatrix}\right)$. Then $\mathfrak{A}$ preserves (ϱ,σ) and $B = \{00, 01, 02, 20\}$ is a ϱ-basis of X^2. The impulse responses are $\lambda(00) = 00$, $\lambda(01) = 01$, $\lambda(10) = 11$, $\lambda(02) = 02$, $\lambda(20) = 22$ ($\lambda(x_1x_2)$ denotes here $\lambda(z_0,x_1x_2)$). Taking into account that $\mathfrak{A}$ preserves (ϱ,σ) we can compute $\lambda(11)$, $\lambda(12)$ and $\lambda(22)$.
As the first letters of 11 and 10 coincide, the first letters of $\lambda(10) = 11$ and $\lambda(11)$ coincide, too, i. e. $\lambda(11) = 1x$ with $x \in \{0,1,2\}$

Since $\begin{pmatrix}(11)\\(00)\end{pmatrix} \subseteq \varrho$, we have $\begin{pmatrix}(11)\\(00)\end{pmatrix} \subseteq \sigma$ (because $\mathfrak{A}$ preserves (ϱ,σ)), thus $\begin{pmatrix}x\\0\end{pmatrix} \subseteq \sigma$, i. e. $x \in \{1,2\}$. Analogously we get $\lambda(12) = 12$, $\lambda(21) = 22$, $\lambda(22) = 22$.

In this case we even can exactly determine the full input-output- behaviour (on X^2) because

$$\begin{pmatrix}(22)\\(11)\end{pmatrix} \subseteq \varrho \quad \text{implies} \quad \begin{pmatrix}\lambda(22)\\\lambda(11)\end{pmatrix} = \begin{pmatrix}(22)\\(1x)\end{pmatrix} \subseteq \sigma$$

consequently $x = 1$, i. e. $\lambda(11) = 11$.

The theory of invariant relations is at the very beginning. However, in future it could be an effective tool to deal with nonlinear systems.

Appendix I

Homomorphic systems

This appendix deals with such nonlinear systems that can be interpreted as linear transformations of vector spaces under an appropriate definition of vector spaces.
The elements of the vector spaces are (infinite) sequences in the time-discrete case resp. time functions in the time-continuous case.

These sequences or time functions are interpreted as the input and output sequences or functions of the systems under consideration. Such systems introduced in /31/ are called homomorphic systems. Often it is realistic to assume the input- and output-vector spaces to be separable Hilbert spaces. Then it can be shown that there exists a canonical representation of homomorphic systems. Homomorphic systems are applied especially in seismic data processing and other areas like speech- and EEG-analysis.

As we have mentioned, we consider the input- and output-sequences (or the input- and output-functions) of a system S as elements of the vector spaces U and V, respectively, over the field F. Formally a system S can be described by a mapping $\varphi_S : U \rightarrow V$. A system S is called invertible if φ_S is invertible.

Let U be a vector space under the vector addition $\oplus$ and the scalar multiplication $\odot$, and let V be a vector space under the vector addition $\boxplus$ and the scalar multiplication $\boxdot$. Then S is called homomorphic if φ_S is homomorphic, i. e. if we have

$$\varphi(c_1 \odot u_1 \oplus c_2 \odot u_2) = c_1 \boxdot \varphi(u_1) \boxplus c_2 \boxdot \varphi(u_2),$$
$$u_1, u_2 \in U, \quad c_1, c_2 \in F.$$

It is easy to see that every invertible system S is homomorphic if the domain of φ_S constitutes a vector space.

<u>Theorem 1.</u> Let S be an invertible system with the input-vector space U under the vector addition $\oplus$ and the scalar multiplication $\odot$, and with outputs in V. Then V constitutes a vector space under the vector addition $\boxplus$ and scalar multiplication $\boxdot$ defined by

$$v_1 \boxplus v_2 = \varphi_S(\varphi_S^{-1}(v_1) \oplus \varphi_S^{-1}(v_2)) \qquad v_1, v_2 \in V$$

$$c \boxdot v = \varphi_S(c \odot \varphi_S^{-1}(v)) \qquad c \in F_u, \quad v \in V$$

and S is homomorphic.

Example. Let U be the set of real functions $f : \mathbb{R} \rightarrow \mathbb{R}$ with the ordinary addition + and the scalar multiplication ordinary multiplication, and let $\varphi(f(t)) = e^{f(t)}$, $\varphi^{-1}(g(t)) = \log g(t)$. Then $\boxplus$ and $\boxdot$ are defined by

$$g_1(t) \boxplus g_2(t) = \varphi(\varphi^{-1}(g_1(t)) + \varphi^{-1}(g_2(t)) = e^{\log g_1(t) + \log g_2(t)} = g_1(t) \cdot g_2(t) \qquad (1)$$

$$c \boxdot g(t) = \varphi(c\,\varphi^{-1}(g(t)) = e^{c \log g(t)} = (g(t))^c \qquad (2)$$

Since $e^x > 0$ for $x \in \mathbb{R}$, the log-function in Eqs. (1) and (2) is well-defined. The function φ satisfies the equation

$$\varphi(c_1 f_1(t) + c_2 f_2(t)) = c_1 \boxdot \varphi(f_1(t)) \boxplus c_2 \boxdot \varphi(f_2(t)) = \varphi(f_1(t))^{c_1} \cdot \varphi(f_2(t))^{c_2} \qquad (3)$$

and φ is homomorphic.

Now we are interested in non-invertible homomorphic systems. A homomorphic mapping $\varphi : U \rightarrow V$ induces an invertible mapping $\tilde{\varphi} : U/N \rightarrow V$, where N is the null-space of φ. This fact allows us to investigate invertible mappings defined on the set of cosets U/N instead of non-invertible ones defined on the vector space U.
A characterization of non-invertible homomorphic systems is given in the next theorem.

Theorem 2. Let S be a non-invertible system with the input-vector space U under the vector addition $\oplus$ and the scalar multiplication $\odot$. Let $V = \{\varphi_S(u) \,/\, u \in U\}$ be the set of outputs of S and let $\{U_u \,/\, u \in U\}$ be the set of equivalence classes defined by

$$u, u' \in U_u \quad \text{if and only if} \quad \varphi_S(u) = \varphi_S(u').$$

Then S is homomorphic if and only if

1. U_o is a vector space, where 0 is the null element of U,

and

2. the set of equivalence classes $\{U_u \,/\, u \in U\}$ equals the set of cosets $\{u \oplus U_o \,/\, u \in U\}$.

If we assume the input- and output-vector spaces U and V, respectively, of a homomorphic system S to be separable Hilbert spaces, then it is easy to derive a canonical representation of S.

Every separable Hilbert space is isomorphic to the space l^2, which constitutes a vector space under ordinary addition and ordinary scalar multiplication.

l^2 denotes the set of all infinite sequences $a_1\ a_2\ a_3\ \ldots,\ a_i \in \mathbb{R}$, with the condition

$$\sum_{i=1}^{\infty} a_i^2 < \infty .$$

Let $a = a_1\ a_2\ a_3\ \ldots$ and $b = b_1\ b_2\ b_3\ \ldots$

Then the sum $a + b$ is defined by

$$a + b = (a_1 + b_1)(a_2 + b_2)(a_3 + b_3)\ \ldots$$

The $\underline{0}$-vector is $0\ 0\ 0\ \ldots$, and $-a$ is $= -a_1\ -a_2\ -a_3\ \ldots$

The inner product (a,b) is defined by

$$(a,b) = \sum_{i=1}^{\infty} a_i b_i .$$

Since U is supposed to be a separable Hilbert space, an orthonormal basis $\{\mathfrak{X}_i\}$ has to exist and every $u \in U$ can be uniquely represented as

$$u = \sum_{i=1}^{\infty} (u, \mathfrak{X}_i)\, \mathfrak{X}_i, \tag{4}$$

from which we obtain

$$\sum_{i=1}^{\infty} (u, \mathfrak{X}_i)^2 = (u,u) < \infty .$$

Therefore $(u, \mathfrak{X}_1)(u, \mathfrak{X}_2)(u, \mathfrak{X}_3)\ \ldots$ is an element of l^2.

We define

$$\gamma(u) = (u, \mathfrak{X}_1)(u, \mathfrak{X}_2)(u, \mathfrak{X}_3)\ldots \tag{5}$$

Since the representation of u by Eq. (4) is unique, γ is invertible.

Let now S be a homomorphic system with the input- and output-vector spaces U and V, respectively, and let U and V be separable Hilbert spaces.

Then there exist isomorphisms $\alpha : U \to l^2$ and $\beta : V \to l^2$, and the mapping $\varphi_S : U \to V$ can be realized as a serial connection of

$\alpha - \alpha^{-1} - \varphi_S - \beta - \beta^{-1} = \varphi_S$. The function $\alpha^{-1} - \varphi_S - \beta : l^2 \to l^2$ transforms l^2 into l^2, which constitutes linear vector spaces under ordinary addition. Therefore $\alpha^{-1} - \varphi_S - \alpha$ is a linear transformation and the corresponding system $L_{\alpha\beta}$ is linear.
These facts are illustrated in Fig. 1.

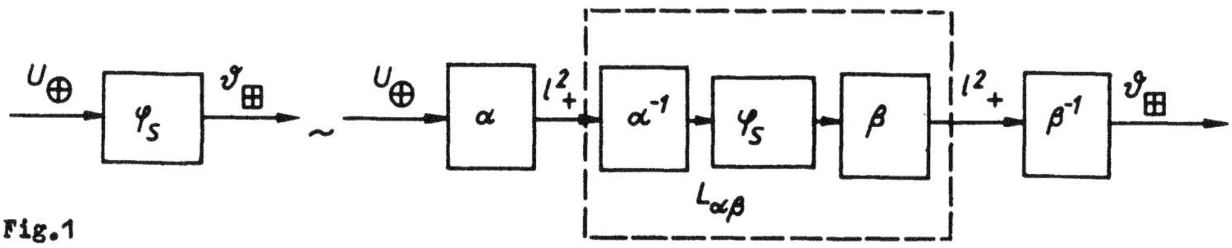

Fig.1

The isomorphisms α and β are determined by U and V, respectively. Thus, for given vector spaces U and V different homomorphic systems differ by their linear parts only.

In general the mapping $\gamma : U \to l^2$ defined by Eq. (5) is not a causal one. If, for instance, $u \in U$ is a continuous time-function, the inner product $(u, \mathfrak{X}_i)$ is determined by

$$(u, \mathfrak{X}_i) = \int_{-\infty}^{\infty} u(\tau)\, \mathfrak{X}_i(\tau)\, d\tau .$$

To compute $(u, \mathfrak{X}_i) \in l^2$ we have to know u(t) in the whole interval $(-\infty, \infty)$. If we wanted to have a physically possible realization of the mapping γ, we would have to consider causal mappings only (s. e. g. /46/). Special causal mappings are so-called memoryless mappings.
Thus the question arises whether the isomorphisms α or β are memoryless or not.
Thereby α is called memoryless if the output at any instant of time depends only on the input at the same instant of time. The isomorphism α can be chosen memoryless if and only if the operations $\oplus$ and $\odot$ in U are memoryless.
Let us denote the value of $u \in U$ at time t by u_t. Then $\oplus$ and $\odot$ are memoryless if

$$(u_1 \oplus u_2)_t = u_{1t} \oplus_t u_{2t}$$

and

$$(c \odot u)_t = c \odot_t u_t .$$

The associativity of $\oplus$ implies the associativity of $\oplus_t$.
Let u_t, $u \in U$, $t > 0$, be real numbers. If $\oplus_t$ is a continuous operation, $u_{1t} \oplus_t u_{2t}$ can be expressed as

$$u_{1t} \oplus_t u_{2t} = \gamma_t(\gamma_t^{-1}(u_{1t}) + \gamma_t^{-1}(u_{2t})),$$

where γ_t is a strictly monotonic function (cf. chapter 3 Eq. (77)).

Homomorphic systems have an important application in signal processing if the signals under consideration are combined by convolution.
Let us consider sequences $x(0)\ x(1)\ x(2)\ \ldots$ with $x(0) \neq 0$ and $\sum_{i=0}^{\infty} x(i)^2 < \infty$ as elements of a vector space. Let the vector addition

$$x(0)x(1)x(2)\ \ldots = x_1(0)x_1(1)x_1(2)\ \ldots \quad x_2(0)x_2(1)x_2(2)\ \ldots$$

be defined by a convolution, i. e.

$$x(n) = \sum_{i=0}^{n} x_1(n-i)\, x_2(i) = \sum_{i=0}^{n} x_1(i)x_2(n-i) \tag{6}$$

for $n > 0$.
As usual the Z-transform of a sequence $x(1)x(2)x(3)\ldots$ is defined by

$$Z[x] = \sum_{k=0}^{\infty} x(k)\ z^{-k} \tag{7}$$

z is a complex variable. $Z[x]$ is a power series in z^{-1} which converges for $|z| > R$, where R is some positive real number. By definition $x(0)x(1)x(2)\ldots$ is the inverse Z^{-1} of the Z-transform $Z[x]$.

Since the Z-transform of a convolution is the product of the Z-transforms, we obtain from Eq. (6)

$$Z[x] = Z[x_1] \cdot Z[x_2] \tag{8}$$

and therefore

$$\log Z[x] = \log Z[x_1] + \log Z[x_2], \tag{9}$$

where log is the complex logarithm.
By application of Z^{-1} to Eq. (47) we get

$$Z^{-1}(\log Z[x]) = Z^{-1}(\log Z[x_1]) + Z^{-1}(\log Z[x_2]) \tag{10}$$

Each of the terms in Eq. (10) represents a sequence.
Denoting $Z^{-1}(\log Z[x]) = \hat{x} = \hat{x}(0)\hat{x}(1)\hat{x}(2)$

Eq. (10) becomes

$$\hat{x} = \hat{x}_1 + \hat{x}_2$$

and the sequences $\hat{x}_1$ and $\hat{x}_2$ are additively combined.
To separate $\hat{x}_1$ and $\hat{x}_2$, a linear filter, i. e. a linear automaton $\mathcal{L}$, can be applied.
Thus the transformations α and α^{-1} are determined by

$$\alpha(x) = Z^{-1}(\log(Z[x])) = \hat{x}$$

$$\alpha^{-1}(\hat{x}) = Z^{-1}(\exp(Z[\hat{x}])) = x$$

The corresponding homomorphic $(\otimes, \otimes)$-superponable system, where $\otimes$ denotes the convolution of sequences, is represented in Fig. 2.

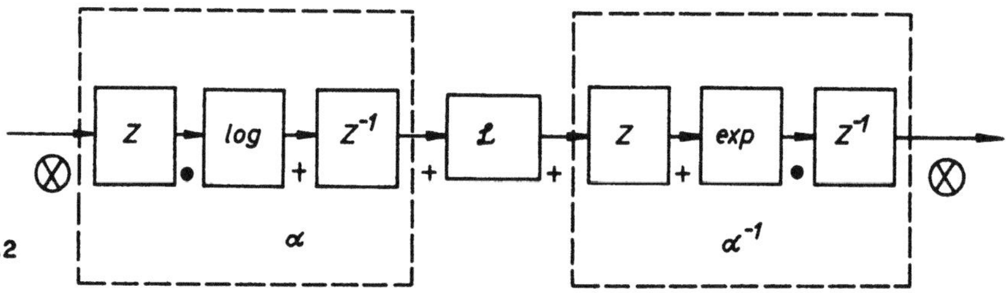

Fig.2

Some care is needed to ensure that the transformations log and Z^{-1} are well-defined. We do not discuss details here.
The first application was in the area of speech analysis /33/. Other applications concern EEG analysis and especially seismic signal processing (s. e. g. /43/, /44/).

Superponable differential equations

Although we are mainly interested in time-discrete systems, we introduce some results concerning the nonlinear superposition of solutions of differential equations in this appendix. We describe the concept of a connecting function of a differential equation and present some known results about connecting functions of some special differential equations.

It is well-known that a suitably chosen linear combination of solutions of a linear differential equation is again a solution of this equation. Moreover, the general solution of an ordinary linear differential equation can be expressed as a linear combination of some special or particular solutions (s. e. g. /42/).

A similar result is in force for the nonlinear Riccati-equation

$$y' = f(x) \cdot y^2 + g(x) \cdot y + h(x). \tag{1}$$

If three special solutions y_1, y_2, y_3 of Eq. (1) are known, the general solution y of Eq. (1) can be expressed as

$$y = F_C(y_1, y_2, y_3) = \frac{y_2(y_3 - y_2) - Cy_1(y_3 - y_1)}{y_3 - y_2 - C(y_3 - y_1)}, \tag{2}$$

where C is an arbitrary real constant (s. e. g. /42/).

The nonlinear function $F_C(y_1, y_2, y_3)$ is a solution of the Riccati-equation if y_1, y_2, y_3 are solutions of this equation, i. e. a nonlinear superposition principle is valid for this equation.

The general concept of nonlinear superposition for ordinary and partial differential equations was introduced in /20/.

First we shall deal with an ordinary differential equation of order n, n > 1,

$$y^{(n)} = f(y^{(n-1)}, \ldots, y', y, x) \tag{3}$$

where $y^{(j)}$ denotes the jth derivation of the function y(x).

<u>Definition 1.</u> Let $y = F(y_1, \ldots, y_m, x)$. Then F is called a connecting function of order m, m ≥ 1, for the differential equation (3) if the following condition is fulfilled:

Whenever $y_1, \ldots, y_m$ are solutions of Eq. (3), then $y = F(y_1, \ldots, y_m, x)$ is also a solution of Eq. (3).

From this definition it is obvious that the existence of a nonlinear connecting function implies the possibility of nonlinear superposition of solutions of a differential equation.

Thereby a function $y_i(x)$ is said to be a solution of Eq. (3) if $y_i^{(n)} = f(y_i^{(n-1)},\dots,y_i',y_i x)$.

Solutions $y_1,\dots,y_m,y$ of Eq. (3) are not restricted by any supplementary conditions which are to be satisfied for special values of x by $y_1,\dots,y_m,y$ and their first (n - 1) derivations.

In the case of partial differential equations we formulate the nonlinear superposition principle for functions of two variables only. However, the following definition can be easily extended to cases with more independent variables.

To avoid a too complicated notation, we describe a general partial differential equation for the function u(x,y) in a compact form as

$$\phi(u,x,y) = 0 \qquad (4)$$

<u>Definition 1'.</u> Let $u = F(u_1,\dots,u_m,x,y)$. Then F is called a connecting function of order m, $m > 1$, for the partial differential equation (4) if the following condition is fulfilled.

Whenever $u_1,\dots,u_m$ are solutions of Eq. (4), then $u = F(u_1,\dots,u_m,x,y)$ is also a solution of Eq. (4).

In general there exist different connecting functions for a given differential equation.

As an example we consider the partial differential equation

$$u_x + u_y = u. \qquad (5)$$

It is easy to see that for any real number n, $n \neq 0$,

$$F_n(u_1,u_2) = (u_1^n + u_2^n)^{\frac{1}{n}} \qquad (6)$$

is a connecting function of Eq. (5) (cf. /20/).

If $n \neq 1$, $F_n(u_1,u_2)$ is a nonlinear connecting function for a linear partial differential equation.

A class of differential equations for which the only possible connecting functions are linear ones is characterized in /40/.

For the differential equation

$$y^n = f(y^{n-1},\dots,y'.y.x), \quad n \geq 2 \qquad (3)$$

the function $M(x,y,\eta)$ is defined by

$$M(x,y,\eta) = \max (u_{n-1},\dots,u_1,y,x)$$
$$u_1 \le \eta \,,\; u_2 \le \eta^2,\dots,u_{n-1} \le \eta^{n-1}. \tag{7}$$

It is assumed that the function M satisfies

$$\lim_{\eta\to\infty} \eta^{-n} M(x,y,\eta) = 0. \tag{8}$$

Condition (8) is (for instance) satisfied if $f(u_{n-1},\dots,u_1,y,x)$ is a bounded function of $u_1,\dots,u_{n-1}$ for any fixed y and x. If

$$y^{(n)} = \sum_{i=0}^{n-1} f_i(x,y) \cdot (y^{(i)})^{k_i}, \tag{9}$$

we get from (8) for $i = 1,\dots,n-1$ the condition

$$i \cdot k_i < n. \tag{10}$$

Thus, for instance

$$y'' = -y'^2 \tag{11}$$

fails to satisfy Condition (8).
For this class of ordinary differential equations of at least second order the following theorem can be proved /40/.

<u>Theorem 1.</u> Let F be a connecting function of order k, $k \ge 1$, for the differential equation (3). If M defined by Eq. (7) satisfies condition (8), F is linear, i. e.

$$F(y_1,\dots,y_k,x) = \gamma(x) + \sum_{i=1}^{k} \alpha_i(x) y_i. \tag{12}$$

$\gamma(x)$, $\alpha_1(x),\dots,\alpha_k(x)$ do not depend on $y_1,\dots,y_k$.

Other theorems concerning linear connecting functions with constant coefficients are also contained in /40/.
Nevertheless, differential equations which fail to satisfy Condition (8) may have nonlinear connecting functions.
For instance

$$F(y_1,y_2) = \log(e^{y_1} + e^{y_2}) \tag{13}$$

is a nonlinear connecting function for Eq. (11).

A method for deriving necessary (and sufficient) conditions for the existence of a connecting function is presented in /25/.
The following theorem is formulated for $n \geq 2$ and proved for $n = 2$.

Theorem 2. Let $y^{(n)} - f(y^{(n-1)},\dots,y',y,x) = 0$ (3)
be a differential equation of order n, $n \geq 1$. Then $F(y_1,\dots,y_m,x)$, $n \geq m > 1$, is a connecting function of Eq. (25) if and only if there exists a function $g(y,x)$ such that

$$y^{(n)} - f(y^{(n-1)},\dots,y',y,x) = \frac{1}{g_y(y,x)} \left\{ \sum_{k=1}^{n} a_k(x) \cdot \frac{d^k}{(dx)^k} g(y,x) + a_o(x) \cdot g(y,x) + b(x) \right\} \tag{14}$$

with $a_n(x) = 1$ and F satisfies

$$g(F(y_1,\dots,y_m),x) = \sum_{i=1}^{m} C_i g(y_i,x), \tag{15}$$

where $C_1,\dots,C_m \in \mathbb{R}$ are constants. If $b(x) \neq 0$, these constants are restricted by

$$\sum_{i=1}^{n} C_i = 1. \tag{16}$$

The explicit form of condition (14) for $n = 2$ is

$$y'' - f(y',y,x) = y'' + \frac{g_{yy}}{g_y} y'^2 + \frac{2g_{xy}}{g_y} y' + \frac{g_{xx}}{g_y} + a_1(x) \left\{ y' + \frac{g_x}{g_y} \right\} + a_o(x) \cdot \frac{g}{g_y} + \frac{b(x)}{g_y} . \tag{17}$$

There is a close relation between differential equations for which a nonlinear connecting function exists and linear equations. If we substitute $Y = g(y,x)$, Eq. (14) becomes linear.

Now we discuss some results concerning partial differential equations. The following theorem is proved in /24/.

Theorem 3. Let

$$u_{xx} = f(u_{xy},u_{yy},u_x,u,x,y). \tag{18}$$

A necessary and sufficient condition that a twice differentiable function $F(u_1,u_2)$ is a connecting function of Eq. (18) is:

1. $f(u_{xy},u_{yy},u_x,u_y,u,x,y) = \{a_1u_{xy} + a_2u_{yy} + a_3u_x + a_4u_y\} - - \frac{g''(u)}{g'(u)}\{u_x^2 + a_1u_xu_y + a_2u_y^2\} + a_5\frac{g(u)}{g'(u)} + \frac{a_6}{g'(u)}$ (19)

   where $a_1,\dots,a_6$ are functions of x and y; g is a twice differentiable function.

2. $F(u_1,u_2)$ is a solution of

   $g(F(u_1,u_2)) = \beta_1\, g(u_1) + \beta_2\, g(u_2)$

   where $\beta_1, \beta_2 \in \mathbb{R}$, with $\beta_1 + \beta_2 = 1$ if $a_6 \neq 0$.

3. $F(u_1,u_2)$ is a function for which the chain rule

$$\frac{\partial^2}{\partial x_i\,\partial x_j} F(u_1,u_2) = \sum_{l,k=1}^{2} \frac{\partial^2 F}{\partial u_l\,\partial u_k}\,\frac{\partial u_l}{\partial x_i}\,\frac{\partial u_k}{\partial x_j} + \sum_{k=1}^{2}\frac{\partial F}{\partial u_k}\,\frac{\partial^2 u_k}{\partial x_i\,\partial x_j}$$

$$i,j = 1,2 \qquad (20)$$

   is satisfied ($x = x_1$, $y = x_2$). (If F is an elementary function, Eq. (20) is satisfied.)

As an example let us consider (s. /20/) the equation

$$\frac{1}{\lambda}u_t + \frac{1}{2\lambda}u_x^2 = u_{xx} \quad . \qquad (21)$$

With $g(u) = e^{-\frac{u}{2\lambda}}$, $-\frac{g''(u)}{g'(u)} = \frac{1}{2\lambda}$, $a_1 = a_2 = a_3 = a_5 = a_6 = 0$ $a_4 = \frac{1}{\lambda}$

the condition (19) is satisfied and a connecting function of Eq. (21) is

$$F(u_1,u_2) = -2\lambda\,\log(\beta_1 \cdot e^{-\frac{u_1}{2\lambda}} + \beta_2 \cdot e^{-\frac{u_2}{2\lambda}}) \qquad (22)$$

for $\beta_1, \beta_2 \in \mathbb{R}$.

In /20/ connecting functions of Eq. (89) are used to derive connecting functions of the equation

$$v_t + v \cdot v_x = \lambda v_{xx}. \qquad (23)$$

Differentiating Eq. (21) with respect to x, we obtain

$$u_{xt} + u_xu_{xx} = \lambda u_{xxx}. \qquad (24)$$

Substituting $u_x = v$, Eqs. (23) and (24) coincide. If u is a solution of Eq. (21), then $v = u_x$ is a solution of Eq. (23).
On the other hand, if $v(x,t)$ is a solution of Eq. (23), then

$u(x,t) = \int^{x} v(\xi,t)\, d\xi$ is a solution of (21) since

$$u_t + \frac{1}{2} u_x - \lambda u_{xx} = \int^{x} v_t(\xi,t)\, d\xi + \frac{v^2(x,t)}{2} - \lambda v_x(x,t) =$$

$$= \int^{x} - \frac{d}{d\xi}\left(\frac{v^2(\xi,t)}{2}\right) d\xi + \lambda \int^{x} v_{\xi\xi}(\xi,t)\, d\xi +$$

$$+ \frac{v^2(x,t)}{2} - \lambda v_x(x,t) = 0.$$

This property can be used to find connecting functions of Eq. (23) from connecting functions of Eq. (21).
Thus

$$\tilde{F}(v_1,v_2) =$$

$$= -2\lambda \frac{\partial}{\partial x} \log\left(\beta_1 \exp\left(-\frac{1}{2\lambda} \int^{x} v_1(\xi,t)\, d\xi\right) + \beta_2 \exp\left(-\frac{1}{2\lambda} \int^{x} v_2(\xi,t)\, d\xi\right)\right) \quad (25)$$

is a connecting function of Eq. (23).
Denoting $g(v) = \exp(-\frac{1}{2\lambda} \int^{x} v(\xi,t)\, d\xi)$, $g^{-1}(v) = -2\lambda \frac{\partial}{\partial x} \log v$, $\tilde{F}(v_1,v_2)$ can be expressed as

$$\tilde{F}(v_1,v_2) = g^{-1}(\beta_1 g(v_1) + \beta_2 g(v_2)).$$

The method shortly discussed here is extended in /30/ to the equation

$$\frac{\partial^2 v}{\partial x_n^2} + \sum_{k=1}^{n-1} b_k(x_1,\dots,x_{n-1}) \frac{\partial u}{\partial x_k} + b_n(x_1,\dots,x_n) \frac{\partial v}{\partial x_n} + v \frac{\partial}{\partial x_n} b_n(x_1,\dots,x_n) = 2\mu v \frac{\partial v}{\partial x_n} \quad (26)$$

Eq. (23) is a special case of Eq. (26) with $n = 2$, $x_2 = x$, $x_1 = t$, $b_2 = 0$, $b_1 = \text{const.}$

As another example of an equation for which a connecting function is known let us consider /30/

$$\sum_{i=1}^{n} a_i(x_1,\dots,x_n) \frac{\partial u}{\partial x_i} = c(x_1,\dots,x_n)\, g(u). \quad (27)$$

$F(u_1,\dots,u_m)$ is a connecting function if

$$\sum_{j=1}^{m} g(u_j) \frac{\partial F}{\partial u_j} = g(F). \quad (28)$$

Eq. (28) is a partial differential equation for the connecting function, and solutions of Eq. (28) are in /30/.

Connecting functions of some other partial differential equations can be found in /21, 22/.

At the end of this appendix let us discuss the ordinary differential equation

$$v \cdot v' = \lambda \cdot v'' \tag{29}$$

arising from Eq. (22) if $v_t = 0$ and if we denote $v_x = v'$, $v_{xx} = v''$. Obviously $F(v_1, v_2)$ defined by Eq. (24) is a connecting function of Eq. (29).

On the other hand, the conditions of Theorem 10 for the existence of an elementary connecting function (for which the chain rule of differentiation can be applied) are

$$\frac{g_{yy}}{g_y} = 0 \tag{30}$$

$$\frac{2g_{yx}}{g_y} + a_2(x) = -\frac{1}{\lambda}\, y \tag{31}$$

$$a_2(x)\,\frac{g_x}{g_y} + \frac{g_{xx}}{g_y} + a_3(x)\, g = 0 \tag{32}$$

Eq. (30) implies $g(x,y) = y \cdot f(x) + h(x)$ in contradiction to Eq. (31).

Thus there do not exist elementary connecting functions. This result shows how to enlarge the class of differential equations for which connecting functions can be derived. Roughly speaking we have to consider non-elementary connecting functions.

References

/1/ Aczel, J., Vorlesungen über Funktionalgleichungen und ihre Anwendungen. Birhäuser Verlag, Basel-Stuttgart 1961.

/2/ Alt, W., Über die reellen Funktionen einer reellen Veränderlichen, welche ein rationales Additionstheorem besitzen. Deutsche Mathematik 5 (1940), 1 - 12.

/3/ Arbib, M. A., Tolerance Automata. Kybernetika (Praha) 3 (1967), 223 - 233.

/4/ Arbib, M. A., Automata Theory. In: Topics in Mathematical System Theory (Eds.: R. E. Kalman, P. L. Falb and M. A. Arbib); Mc Graw Hill, New York 1969.

/5/ Bochmann, D., Über Ableitungen in der Schaltalgebra und einen damit formulierbaren Entwicklungssatz. Wiss. Z. TU Dresden, 14 (1965), 1523 - 1527.

/6/ Bondarcuk, W. G., L. A. Kalužnin, W. N. Kotow, B. A. Romow, Galois Theorie für Postsche Algebren I, II. (in russ.) Kibernetika (Kiew) 3/1969, 1 - 10, 5/1969, 1 - 9.

/7/ Föllinger, O., Nichtlineare Regelungen I, II, 2. Aufl. Oldenburg-Verlag, München-Wien 1978.

/8/ Gill, A., Introduction to the Theory of Finite State Machines. Mc Graw Hill, New York 1962.

/9/ Gill, A., Linear Sequential Circuits. Mc Graw Hill, New York 1966.

/10/ Gill, A., The reduced form of a linear automaton. In: Automata Theory (Ed.: E. R. Caianello) Academic Press, New York 1966.

/11/ Gössel, M., Angewandte Automatentheorie II. Akademie-Verlag, Berlin 1971.

/12/ Gössel, M., Binäre +- und ≡-superponierbare Automaten. Z. f. elekt. Inform.- u. Energietechnik (IET) 6 (1976), 225 - 235.

/13/ Gössel, M., Binäre $\vee$- und $\wedge$-superponierbare Automaten. Z. f. elektr. Inform.- u. Energietechnik (IET) 6 (1976), 325 - 330.

/14/ Gössel, M., Zur Verallgemeinerung der Theorie linearer zeitdiskreter Systeme. Z. f. elektr. Inform.- und Energietechnik (IET) 8 (1978), 517 - 522.

/15/ Gössel, M., Generalized superposition for binary automata with respect to a pair of Boolean functions. Kibernetika (Kiew) 3/ 1978, 33 - 37.

/16/ Gössel, M., Über φ-lineare kontinuierliche Systeme. Z. f. elektr. Inform.- u. Energietechnik (IET) 10 (1980), 525 - 533.

/17/ Gössel, M., H. D. Modrow, Verallgemeinerte Superposition bei binären Automaten. Acta Cybernetica (Szeged) 3 (1977), 163 - 172.

/18/ Gössel, M., Pöschel, R., On the characterization of linear and linearizable automata by a superposition principle. Math. Syst. Theory 11 (1977), 61 - 76.

/19/ Gössel, M., R. Pöschel, Invariant Relations for Automata- a Proposal. EIK 16 (1980) 4, 147 - 169.

/20/ Jones, S. E., W. F. Ames, Nonlinear Superposition. J. Math. Analysis and Appl. 17 (1967), 484 - 487.

/21/ Kečkić, J. D., On nonlinear superposition. Math. Balkanika 2 (1972) 88 - 93

/22/ Kečkić, J. D., On nonlinear superposition II. Math. Balkanica 3 (1973), 206 - 212.

/23/ Kocič, V. Lj., Theorems on Nonlinear Superposition, the general first order equation. Univ. Beograd Publ. Elektrotechn. Fak. Ser. Math. Fiz. No 577 - 598, (1977), 45 - 48.

/24/ Kocič, V. Lj., Theorems on Nonlinear Superposition II. The general second order equation. Univ. Beograd Publ. Elektrotechn. Fak. Ser. Math. Fiz. No 634 - 677, (1979), 108 - 115.

/25/ Kocič, V. Lj., Theorems on Nonlinear Superposition III, Ordinary Differential Equations. Publ. Inst. Math. Beograd 25 (39), (1979), 55 - 62.

/26/ Kohavi, Zv., Switching and Finite Automata Theory. Mc Graw Hill, New York 1970.

/27/ Kulik, L. A., Zeitvariable superponierbare digitale Systeme. Diss. A, TU Dresden, 1976.

/28/ Kulik, L. A., Zeitvariable superponierbare digitale Systeme - ein Beitrag zur Analyse und Synthese. Wiss. Z. TU Dresden, 27 (1978), 1181 - 1182.

/29/ Kuwagaki, A., Sur l'equation functionelle: f(x + y) = R(f(x),f(y)). Mem. of the Coll. of Sci. Univ. Kyoto, Series A. 26 (1950), 139 - 144.

/30/ Levin, S. A., Principles of Nonlinear Superposition. J. Math. Analysis and Appl. 30 (1970), 197 - 205.

/31/ Oppenheim, A. V., Superposition in a class of nonlinear systems. Techn. Report No. 432, Research Lab. Electronics, MIT, Cambridge (Mass.) 1965.

/32/ Oppenheim, A. V., Generalized Superposition. Inform. and Control 11 (1967), 528 - 536.

/33/ Oppenheim, A. V., W. Schafer, Homomorphic Analysis of Speech. IEEE Trans. Audio and Electroacustics AU-16 (1968), 221 - 226.

/34/ Pöschel, R., M. Gössel, Generalization of the Theory of Linear Automata. In: IFAC-Symposium Diskrete Systeme, Dresden 1977, Heft 5, S. 65 - 73.

/35/ Pöschel, R., L. A. Kalužnin, Funktionen- und Relationenalgebren. VEB Deutscher Verlag der Wissenschaften, Berlin 1979.

/36/ Reischer, C., D. A. Simovici, Linear Boolean Automata. Wiss. Z. TH Ilmenau 17 (1971), 83 - 96.

/37/ Reusch, B., Theorems on additive automata. In: Theory of Machines and Computation (Eds.: Zv. Kohavi, A. Paz). Academic Press, New York 1971, pp. 243 - 253.

/38/ Rosenberg, I., Über die funktionale Vollständigkeit in den mehrwertigen Logiken. Rozpravy Československé Akad. Věd. 80 (1970)

/39/ Shegalkin, I. I., Arithmetrisierung der sysmbolischen Logik (in russ.). Mathem. Sbornik 35 (1928), 311 - 377.

/40/ Spijker, M. N., Superposition in Linear and Nonlinear Ordinary Differential Equations. J. Math. Analysis and Appl. 30 (1970), 206 - 222.

/41/ Steinmüller, K. H., M. Gössel, (o,$\triangledown$)-linear automata with rational operations o and $\triangledown$. Found. Control Eng. (Poznan) 3 (1978), 41 - 48.

/42/ Stepanow, W. W., Lehrbuch der Differentialgleichungen. VEB Deutscher Verlag der Wissenschaften, Berlin 1967 (Übers. aus d.

Russ.).

/43/ Tribolet, J. M., Seismic Applications of Homomorphic Signal Processing. Prentice Hall, Englewood Cliffs (New Jersey) 1979.

/44/ Ulrych, T. J., Application of Homomorphic Deconvolution to Seismology. Geophysics 36 (1971), 650 - 660.

/45/ Wulich, B. S., Einführung in die Funktionalanalysis I. Teubner Verlag, Leipzig 1961 (transl. from the russ.).

/46/ Wunsch, G., Moderne Systemtheorie. Geest & Portig, Leipzig 1962.

Index

Lecture Notes in Control and Information Sciences

Edited by A. V. Balakrishnan and M. Thoma

Vol. 22: Optimization Techniques
Proceedings of the 9th IFIP Conference on Optimization Techniques,
Warsaw, September 4–8, 1979
Part 1
Edited by K. Iracki, K. Malanowski, S. Walukiewicz
XVI, 569 pages. 1980

Vol. 23: Optimization Techniques
Proceedings of the 9th IFIP Conference on Optimization Techniques,
Warsaw, September 4-8, 1979
Part 2
Edited by K. Iracki, K. Malanowski, S. Walukiewicz
XV, 621 pages. 1980

Vol. 24: Methods and Applications in Adaptive Control
Proceedings of an International Symposium
Bochum, 1980
Edited by H. Unbehauen
VI, 309 pages. 1980

Vol. 25: Stochastic Differential Systems – Filtering and Control
Proceedings of the IFIP-WG7/1 Working Conference
Vilnius, Lithuania, USSR, Aug. 28 – Sept. 2, 1978
Edited by B. Grigelionis
X, 362 pages. 1980

Vol. 26: D. L. Iglehart, G. S. Shedler
Regenerative Simulation of Response Times in Networks of Queues
XII, 204 pages. 1980

Vol. 27: D. H. Jacobson, D. H. Martin, M. Pachter, T. Geveci
Extensions of Linear-Quadratic Control Theory
XI, 288 pages. 1980

Vol. 28: Analysis and Optimization of Systems
Proceedings of the Fourth International Conference on Analysis and Optimization of Systems
Versailles, December 16–19, 1980
Edited by A. Bensoussan and J. L. Lions
XIV, 999 pages. 1980

Vol. 29: M. Vidyasagar,
Input-Output Analysis of Large-Scale Interconnected Systems – Decomposition, Well-Posedness and Stability
VI, 221 pages. 1981

Vol. 30: Optimization and Optimal Control
Proceedings of a Conference Held at Oberwolfach, March 16–22, 1980
Edited by A. Auslender, W. Oettli, and J. Stoer
VIII, 254 pages. 1981

Vol. 31: Berc Rustem
Projection Methods in Constrained Optimisation and Applications to Optimal Policy Decisions
XV, 315 pages. 1981

Vol. 32: Tsuyoshi Matsuo,
Realization Theory of Continuous-Time Dynamical Systems
VI, 329 pages, 1981

Vol. 33: Peter Dransfield
Hydraulic Control Systems – Design and Analysis of Their Dynamics
VII, 227 pages, 1981

Vol. 34: H.W. Knobloch
Higher Order Necessary Conditions in Optimal Control Theory
V, 173 pages, 1981

Vol. 35: Global Modelling
Proceedings of the IFIP-WG 7/1 Working Conference Dubrovnik, Yugoslavia, Sept. 1–5, 1980
Edited by S. Krčevinac
VIII, 232 pages, 1981

Vol. 36: Stochastic Differential Systems
Proceedings of the 3rd IFIP-WG 7/1 Working Conference
Visegrád, Hungary, Sept. 15–20, 1980
Edited by M. Arató, D. Vermes, A.V. Balakrishnan
VI, 238 pages, 1981

Vol. 37: Rüdiger Schmidt
Advances in Nonlinear Parameter Optimization
VI, 159 pages, 1982

Vol. 38: System Modeling and Optimization
Proceedings of the 10th IFIP Conference
New York City, USA, Aug. 31 – Sept. 4, 1981
Edited by R.F. Drenick and F. Kozin
XI, 894 pages. 1982

Vol. 39: Feedback Control of Linear and Nonlinear Systems
Proceedings of the Joint Workshop on Feedback and Synthesis of Linear and Nonlinear Systems
Bielefeld/Rom
XIII, 284 pages. 1982

Vol. 40: Y.S. Hung, A.G.J. MacFarlane
A Quasi-Classical Approach to Multivariable Feedback
X, 182 pages. 1982

Vol. 41: M. Gössel
Nonlinear Time-Discrete Systems – A General Approach by Nonlinear Superposition
VIII, 112 pages. 1982

GPSR Compliance

The European Union's (EU) General Product Safety Regulation (GPSR) is a set of rules that requires consumer products to be safe and our obligations to ensure this.

If you have any concerns about our products, you can contact us on ProductSafety@springernature.com

In case Publisher is established outside the EU, the EU authorized representative is:

Springer Nature Customer Service Center GmbH
Europaplatz 3
69115 Heidelberg, Germany

Batch number: 10683525

Printed by Printforce, the Netherlands